Green Energy and Technology

Climate change, environmental impact and the limited natural resources urge scientific research and novel technical solutions. The monograph series Green Energy and Technology serves as a publishing platform for scientific and technological approaches to "green"—i.e. environmentally friendly and sustainable—technologies. While a focus lies on energy and power supply, it also covers "green" solutions in industrial engineering and engineering design. Green Energy and Technology addresses researchers, advanced students, technical consultants as well as decision makers in industries and politics. Hence, the level of presentation spans from instructional to highly technical.

Indexed in Scopus.

More information about this series at http://www.springer.com/series/8059

Djamila Rekioua

Hybrid Renewable Energy Systems

Optimization and Power Management Control

 Springer

Djamila Rekioua
L.T.I.I Laboratory
University of Bejaia
Bejaia, Algeria

ISSN 1865-3529   ISSN 1865-3537   (electronic)
Green Energy and Technology
ISBN 978-3-030-34023-0   ISBN 978-3-030-34021-6   (eBook)
https://doi.org/10.1007/978-3-030-34021-6

This Springer imprint is published by the registered company Springer Nature Switzerland AG
The registered company address is: Gewerbestrasse 11, 6330 Cham, Switzerland

Acknowledgements

Generally, a book cannot be written without assistance. So therefore, I must thank all the many people who helped me. I would particularly extend special thanks to my colleagues and my Ph.D. students in renewable energy and power electronics teams of Laboratory LT2I (University of Bejaia) for cooperation in common projects in renewable energy.

To my family.

In addition, thanks are due to Springer-Verlag for publishing this book.

Djamila Rekioua

Introduction

Hybrid power systems (HPS) combine two or more sources of renewable energy as one or more conventional energy sources. Renewable energy sources such as photovoltaic and wind do not deliver a constant power, but through their complementarity, their combination provides a continuous electrical output.

The purpose of a hybrid power system is to produce as much energy as possible from renewable energy sources to ensure the load demand. In addition to sources of energy, a hybrid system may also incorporate a distribution system, a storage system, converters, fillers and load management control or supervision system. All these components can be connected in different architectures. The renewable energy sources can be connected to the DC bus depending on the size of the system. The power delivered by a HPS can range from a few watts for domestic applications to a few megawatts for systems used in the electrification of small villages.

Aims of the Book

Many books currently on the market are devoted to the subject of hybrid renewable energy systems. This book covers not only an overview of the different configurations and classification of hybrid systems, but also optimization, storage and sizing with applications under MATLAB/Simulink.

The main objective of this book is to enable all graduate and postgraduate students in the fields of electrical engineering, as well as researchers of renewable energy and researchers of modern power systems to quickly understand the supervision of hybrid systems and provide models for control, optimization and storage. In the first part, we present an overview on the different configurations of hybrid renewable energy systems. And in the second part, we give some applications. Mathematical models are given for each system, and a corresponding example under MATLAB/Simulink package is given at the end of each section. Various examples are given for an eventual implementation under DSPACE package or real-time implementation. In order to optimize wind and photovoltaic systems, intelligent techniques are developed. By focusing on the different configurations of hybrid renewable energy systems, this book provides those involved in research and work in the field of renewable energy solutions with an extensive knowledge of the control, optimization and supervision strategies for the different renewable energy systems. By writing this book, we complement existing knowledge in the field of renewable energy systems and the reader will learn how to make the power management, the supervision and the optimization of the most used hybrid systems and hybrid storage.

How the Book is Organized?

The book is organized through six chapters as follows:

- Chapter 1 is intended as an introduction to the subject. It defines the different hybrid renewable energy systems, introduces the main advantages and disadvantages, the choice of the configurations and the common bus type, and presents an overview of the most used configurations of hybrid systems.
- Chapter 2 focuses on the different structures of converters used in photovoltaic and wind systems. Some applications are given under MATLAB/Simulink.
- In Chap. 3, a detailed review on the most used algorithms to track the maximum power point is presented. Several maximum power point tracking (MPPT) algorithms have been developed to track the maximum power point of the wind turbine and photovoltaic generators. The most used are presented (conventional and advanced ones). Some simple MATLAB/Simulink examples are given.
- In Chap. 4, an overview the storage device is showed. The study describes the different storage used in wind and photovoltaic systems. Also, an attention is given to hybrid storage as super-capacitors/fuel cells, batteries/super-capacitors, batteries/fuel cells, etc., used in electric vehicle. Some energy storage applications under MATLAB/Simulink are presented.
- Chapter 5 focuses on sizing of hybrid renewable energy systems. The different methods used in photovoltaic, wind, fuel cells and hybrid systems are presented. HOMER Pro software is used to size and optimize different hybrid systems. An application is given.
- Chapter 6 is devoted to power management of hybrid wind systems. The chapter describes the different configurations and the different combinations of hybrid systems. Power management controls under different scenarios are presented. Applications, made under MATLAB/Simulink, to standalone systems and electric vehicle are presented.

September 2019   Djamila Rekioua

Contents

1 Hybrid Renewable Energy Systems Overview 1
1.1   Introduction 1
1.2   Advantages and Disadvantages of an Hybrid System 1
1.3   Configuration of Hybrid System 2
      1.3.1   Choice of Common Bus Type.................... 2
      1.3.2   Choice of Converters 5
1.4   Classifications of Hybrid Energy Systems................. 6
1.5   Different Combinations of Hybrid Systems................. 6
      1.5.1   PV System with Battery Storage.................. 10
      1.5.2   PV System/Fuel Cells........................... 10
      1.5.3   PV System/Fuel Cells with Battery Storage 14
      1.5.4   PV System/FC Multi-storage Batteries/Super-
              Capacitors 14
      1.5.5   Hybrid Wind/Photovoltaic System................. 17
      1.5.6   Hybrid Wind/Photovoltaic System with Battery
              Storage 17
      1.5.7   Hybrid Wind/Photovoltaic System with Flywheels
              Storage 19
      1.5.8   Wind Turbine System with Fuel Cells............. 21
      1.5.9   Wind System/Fuel Cells with Battery Storage 21
      1.5.10 Wind System/Fuel Cells with Hybrid Storage Batteries/
              Super-Capacitors 21
      1.5.11 PV System with Diesel Generators 22
      1.5.12 PV System with Diesel Generators with Battery
              Storage 23
      1.5.13 PV System with Wind Turbine System and Diesel
              Generators................................... 24

	1.5.14	PV System with Wind Turbine and Diesel Generators with Battery Storage	24
1.6	Conclusion		24
References			26

2 Power Electronics in Hybrid Renewable Energies Systems **39**
2.1	Introduction		39
2.2	Power Electronic Converters		39
2.3	Power Converters in PV Systems		40
	2.3.1	DC/DC Converters	40
	2.3.2	Application in PV Systems	50
	2.3.3	Application in Wind System	50
	2.3.4	Application in Storage Systems	51
	2.3.5	Application to Hybrid PV/Wind System with Battery Storage	51
	2.3.6	DC/AC Converters	51
2.4	Examples of Applications in PV Systems		59
	2.4.1	DC/AC Conversion	60
2.5	Inverter Control Strategies		68
	2.5.1	Power Control	68
2.6	Applications to Pumping Systems		70
2.7	Application in Wind Turbine Energy Conversion		70
	2.7.1	AC/DC Converters	71
	2.7.2	AC/AC Converters	71
2.8	Examples of Converter Topologies in Hybrid Systems		72
	2.8.1	Hybrid PV/DG/Wind Turbine with Battery Storage	72
	2.8.2	Hybrid PV/Wind System	72
References			76

3 MPPT Methods in Hybrid Renewable Energy Systems **79**
3.1	Introduction to Optimization Algorithms		79
3.2	MPPT for Hybrid System		79
3.3	Survey of Maximum Power Point Tracking (MPPT) Algorithms in PV Systems		79
	3.3.1	Most Used MPPT Algorithms in Photovoltaic Systems	81
	3.3.2	Efficiency of an MPPT Algorithm	110
	3.3.3	Comparison of Different Algorithms	111
	3.3.4	Global Maximum Power Point Tracking (GMPPT) Techniques of Photovoltaic System Under Partial Shading Conditions	112
3.4	MPPT Algorithms in Wind Turbine Systems		118
	3.4.1	Tip Speed Ratio Method (TSR)	119
	3.4.2	Power Signal Feedback (PSF) Method	120

	3.4.3	Optimal Torque Control (OTC)	121
	3.4.4	Hill Climb Searching (HCS) Technique	124
	3.4.5	MPPT Based on Gradient Method (GM)	124
	3.4.6	Hybrid MPPT Technique HP&O/OTC	126
	3.4.7	Fuzzy Logic Controller Technique	126
	3.4.8	Artificial Neural Networks (ANN) Method	127
	3.4.9	Radial Basis Function Network (RBFN)	129
	3.4.10	Adaptive Neuro-fuzzy Inference System (ANFIS)	130
	3.4.11	Comparison Between Different Optimization Methods	132
3.5	Conclusion		135
References			135

4   Storage in Hybrid Renewable Energy Systems 139
4.1	Introduction		139
4.2	Electrochemical Storage		139
	4.2.1	Application in Hybrid PV-Wind Turbine System	142
	4.2.2	Battery Models	142
	4.2.3	Application in Photovoltaic System	142
4.3	Mechanical Storage		148
	4.3.1	Flywheel Electric Energy Storage (FEES)	148
	4.3.2	Application in Wind System	149
	4.3.3	Pumped Hydro Energy Storage (PHES)	152
4.4	Super-Capacitor Energy Storage (SES)		152
4.5	Fuel Cells		152
4.6	Electrical Characteristics of PEM Fuel Cells		154
4.7	Super-Capacitors		157
4.8	Hybrid Storage		159
	4.8.1	FC/SC Hybrid Storage	159
	4.8.2	Application 1: FCS/SCs for EVs	159
	4.8.3	Fuel Cell/Battery System	160
	4.8.4	Application 2: FCs/Batteries for EVs	161
	4.8.5	Application to Wind Turbine System	161
	4.8.6	Super-Capacitors/Batteries System	166
	4.8.7	Application 3: SCs/Batteries for EVs	166
	4.8.8	Application in Wind Turbine System	168
	4.8.9	Multiple Storage	168
4.9	Conclusions		168
References			169

5   Design of Hybrid Renewable Energy Systems 173
5.1	Introduction		173
5.2	Design of Photovoltaic Systems		173
	5.2.1	Determination of the Load Demand of Consumers	173
	5.2.2	Photovoltaic System Design	175

5.3   Design of Wind System 179
       5.3.1   Calculation of Wind Energy 179
       5.3.2   Determination of the Wind Generator Size........... 179
5.4   Sizing of Hybrid Photovoltaic/Wind System 179
5.5   Sizing of Hybrid Photovoltaic/Wind System/Batteries 182
       5.5.1   Battery Design............................... 182
       5.5.2   DC/AC Converter Design...................... 184
5.6   Design of Hybrid Photovoltaic/Wind System/Fuel Cells 184
       5.6.1   Power Calculation 185
       5.6.2   Cell Number and Cell Surface 185
5.7   Application to Water Pumping System.................... 186
5.8   Optimization of Power System Using HOMER Pro Software ... 187
       5.8.1   Introduction to HOMER Pro Software............. 187
       5.8.2   Application to PV/Wind System with Battery Storage ... 189
5.9   Conclusions 193
References 193

6   Power Management and Supervision of Hybrid Renewable Energy
    Systems 197
    6.1   Introduction 197
    6.2   Different Combinations of Hybrid Systems................. 197
    6.3   Photovoltaic/Battery System 197
           6.3.1   Power Management Algorithm for the PV/Battery
                    Installation (First Structure)..................... 199
           6.3.2   Application of the First Structure Under MATLAB/
                    Simulink 200
           6.3.3   Power Management Algorithm for the PV/Battery
                    Installation (Second Structure) 204
    6.4   Photovoltaic/Fuel Cell System 207
           6.4.1   Power Management of Photovoltaic/Fuel Cell System ... 207
           6.4.2   Application Under MATLAB/Simulink 207
    6.5   Photovoltaic/Battery/Fuel Cell System 210
           6.5.1   Power Management of Photovoltaic/Battery/Fuel Cell
                    System 210
           6.5.2   Application Under MATLAB/Simulink 212
    6.6   Wind/Battery System 213
           6.6.1   Power Management of Wind/Battery System 213
           6.6.2   Application Under MATLAB/Simulink 215
    6.7   Photovoltaic/Wind Turbine/Battery Structure 217
           6.7.1   Power Management of Photovoltaic/Wind Turbine/
                    Batteries 217
           6.7.2   Power Management of Photovoltaic/Wind Turbine/
                    Batteries for Pumping Water..................... 219

	6.7.3	Application Under MATLAB/Simulink	222
6.8	Photovoltaic/Wind Turbine/Battery/Diesel Generator System		222
	6.8.1	Power Management of Photovoltaic/Wind Turbine/ Battery/Diesel Generator System	222
	6.8.2	Application Under MATLAB/Simulink	228
6.9	Photovoltaic/Wind Turbine/Batteries/Fuel Cells.		230
	6.9.1	Power Management of Photovoltaic/Wind Turbine/ Batteries/Fuel Cells	230
	6.9.2	Application Under MATLAB/Simulink	232
6.10	PV System with Hybrid Storage (Batteries/Super-Capacitors) . . .		234
	6.10.1	Control of DC Bus.	235
	6.10.2	Power Management of PV System with Battery/SC Storage for Electric Vehicle	236
6.11	PV/FCs with Hybrid Storage (Batteries/SCs)		238
	6.11.1	Supervision of PV/FCs with Hybrid Storage (Batteries/SCs).	238
6.12	Power Management of Fuel Cell/Battery System Supplying Electric Vehicle		239
	6.12.1	First Structure	241
	6.12.2	Second Structure	241
6.13	Conclusions		247
References			247

Notations

A_f	Frontal surface area of the vehicle
A_{FC}	Active cell area
A_{pv}	Size of photovoltaic generator
$A_{pv\text{-}u}$	Unit PV panel area
$A_{pv\text{-}tot}$	Total PV area
A_{wind}	Wind area
$A_{wind\text{-}total}$	Total area of the wind turbine generators
C_d	Aerodynamic drag coefficient
C_{DC}	DC-link capacitance
C_1 and C_2	Capacities of the battery at different discharge rate states
C_{batt}	Battery capacity
C_{10}	Rated capacity
C_P	Power coefficient
D	Duty cycle
D_{energy}	Daily energy
$D_{energy\text{-}total}$	Total daily energy
DOD	Depth of discharge
$e(x)$	Difference between the controlled variable x and its reference $\hat{x}$
$e_{a,b,c}$	Induced f.e.m in the stator phase windings
E_b	Zero-current voltage of the battery charged
E_c	Kinetic energy stored in the flywheel
E_{c0}	Initial energy of the flywheel
E_{cref}	Reference energy for the flywheel
E_{FC}	Voltage per cell
$\overline{E_{pv}}$	Annual average values of PV monthly contribution
$E_{pv,m}$	Monthly energy produced by the system per unit area is denoted
$E_{s\text{-}STC}$	The irradiation value under standard test conditions (STC)
$E_{s\text{-}Worst}$	Value of the monthly average irradiation of the worst month of irradiation
E_T and E_ϕ	Torque and flux errors

$\overline{E_{wind}}$	Annual average values of wind monthly contribution
$E_{wind,m}$	Monthly energy produced by the system per unit area
E_{wind}	Energy produced by wind generator
f_n	Nominal stator frequency
f	Fraction of load
F_{aero}	Aerodynamic drag force
F_{energy}	Full energy
F_r	Total force
f_s	Stator frequency
F_{slope}	Climbing force
f_{ro}	Rolling resistance force constant
F_{tire}	Rolling resistance force
E_{wind}	Energy produced by the wind generator
E_s	Solar radiation on tilted plane module
G	Solar irradiance
h_{sun}	Peak sun hour
$I_{a,b,c}$	Current phase voltages
I_{batt}	Battery's current
I_{load}	Load's current
I_{mpp}	Current maximum point
I_s	Stator current
I_{stack}	FC current
I_{pv}	PV current
I_{10}	Current which corresponds to the operating speed
j	Current density
J_{FESS}	Inertia flywheel
J_{mot}	Inertia of the AC motor
J_t	Inertia of the turbine
J_w	Moment of inertia of the flywheel
h	Height
K	Pump constant
K_{batt}	Battery relay control signal
K_{DG}	Diesel generator relay control signal
K_{pv}	Photovoltaic relay control signal
K_{wind}	Wind turbine relay control signal
K_1, K_2, K_3, K_4	Switches
L	Stator inductance
L_s	Stator leakage inductance
m	Month of the year
m_{EV}	Vehicle total mass
$M_{energy,m}$	Monthly energy required by the load
N	Number of the different appliances
N_{aut}	Number of days of autonomy
N_{batt}	Battery number

N_{FC}	Number of cells in the stack
N_{pv}	Number of photovoltaic panels
$N_{pv\text{-para}}$	PV parallel panels
$N_{pv\text{-serial}}$	Serial PV panels
N_{wind}	Number of wind turbine generator
$P_{a\text{-PV/day}}$	Actual daily power
P_{batt}	Battery power
P_{DG}	Power delivered from the diesel generator
P_{dump}	Dump power
P_{FC}	Maximum electrical FC power
P_{inv}	Input inverter power
P_{hyb}	Hybrid power
P_{Hyd}	Hydraulic power
P_{load}	Power required by the load
P_{mec}	Mechanical power
P_{mpp}	Power maximum point
P_{out}	Output power
P_{p}	Peak power of panels
$P_{p\text{-total}}$	The total peak power
P_{pv}	Power delivered from the photovoltaic generator
P_{ref}	Active power control
P_{Ren}	Power produced by PV and wind systems
P_{SC}	Super-Capacitor power
$P_{tb\text{-max}}$	Maximum power turbine
P_{wind}	Power delivered from the wind turbine
q_{v}	Water flow rate
R	Total resistive force
r_{tire}	Tire radius
R_{batt}	Internal (ohmic) resistance of the battery
R_{0}	Total internal resistance of a fully charged battery
t	Run time per day
T_{em}	Electromechanical torque
T_{j}	Cell temperature
$T_{j\text{-STC}}$	Cell temperature under STC
T_{r}	Load torque
V	Electric vehicle speed
$V_{a,b,c}$	Machine phase voltages
V_{batt}	Battery voltage
$V_{DC\text{-bus}}$	DC bus voltage
V_{ds} and V_{ds}	(d,q) components of the stator voltage
V_{oc}	Open-circuit voltage
$V_{s\alpha}$ and $V_{s\beta}$	(α,β) component of stator voltage
V_{pv}	Photovoltaic voltage
$V_{pv\text{-max}}$	Maximum terminal voltage of the photovoltaic generator

V_{mpp}	Voltage maximum point
$V_{oc\text{-}STC}$	Open-circuit voltage under standard test conditions
v_{wind}	Wind speed
Q	Accumulated ampere-hours divided by full battery capacity

Acronyms

d, q   quantities in d-axis and q-axis
α,β   quantities in α-axis and β-axis
a, b, c   Quantities in phases a, b and c

Superscripts

AC	Alternative current
ACO	Ant colony optimization
AFLC	Adaptive fuzzy logic controller
ANFIS	Adaptive neuro-fuzzy inference system
ANN	Artificial neural networks
Batt	Batteries
BN	Big negative
BP	Big positive
CIEMAT	Centro De Investigaciones Energéticas, Medioambientales Y Tecnológicas
DC	Direct current
DG	Diesel generator
DOD	Depth of discharge
DTC	Direct torque control
EG	Electrical generator
ES	Energy storage
EV	Electric vehicle
FC	Fuel cells
FES	Flywheel energy storage
FLC	Fuzzy logic controller
FOC	Field-oriented control
GM	Gradient method
GMPPT	Global MPPT

H_2	Hydrogen
HCS	Hill-climb searching
HP	Hydropower
HRES	Hybrid renewable energy systems
IncCond	Incremental conductance algorithm
Li-Ion	Lithium ion
MHDTC	Modulated hysteresis direct torque control
MN	Means negative
MP	Means positive
MPO	Modified perturb and observe
MPPT	Maximum power point tracking
MS	Mechanical storage
MSMSS	Multi-source multi-storage systems
NN	Neural network
NPC	Net present cost
OCV	Open-circuit voltage
OTC	Optimal torque control
P&O	Perturb and observe
PEMFC	Proton exchange membrane fuel cell
PHES	Pumped hydro energy storage
PI	Proportional integral
PMC	Power management control
PMSG	Permanent magnet synchronous generator
PSF	Power signal feedback
PSO	Particle swarm optimization
PVG	Photovoltaic generator
PWM	Pulse width modulation
RBFN	Radial basis function network
SAWS	Standalone wind system
SC	Super-Capacitors
SMC	Sliding mode control
SN	Small negative
SOC	State of charge
SP	Small positive
SRCC	Switching ripple correlation control
SSCC	Fractional short-circuit current
TSR	Tip speed ratio
VSI	Voltage source inverter
WT	Wind turbine
Z	Zero

Symbols

α_{sc}	Temperature coefficient of short current
β	Road slope angle
Δt	Period of time
γ	Temperature coefficient
θ_n	Digitized signals
Θ	Position of the rotor
λ_{opt}	Optimal tip speed ratio
ΔT	Heating of the accumulator
ΔP	Net power
η_{batt}	Efficiency of the battery
η_{el}	Electrical efficiency of the whole installation (charge controller, inverter, etc.)
η_{DC}	Distribution circuit
η_{pv}	Photovoltaic efficiency
ω_t	Turbine speed
ω_g	Generator speed
ω_{ref}	Wind turbine reference speed
ω_{opt}	Optimal reference angular velocity
Ω_w	Speed of the flywheel
Ω_{ref}	Reference speed for the flywheel
ρ_{air}	Air density
ρ_{water}	Water density
σ	Leakage coefficient

List of Figures

Fig. 1.1   Configuration of the hybrid system with DC bus   2
Fig. 1.2   Configuration of the hybrid system with AC bus   3
Fig. 1.3   Configuration of the hybrid system with AC bus
            and DC bus   4
Fig. 1.4   Sources and loads supplied by various static converters   5
Fig. 1.5   Representation of some used hybrid systems   7
Fig. 1.6   Photovoltaic system with battery storage   10
Fig. 1.7   Standalone PV system with battery storage powering
            DC and AC loads   10
Fig. 1.8   PV/batteries under MATLAB/Simulink   11
Fig. 1.9   Different blocks of PV/battery system   11
Fig. 1.10   PV system with fuel cells   12
Fig. 1.11   Hybrid photovoltaic/fuel cell block diagram   12
Fig. 1.12   PV/FC system   13
Fig. 1.13   Fuel cell model   13
Fig. 1.14   PV/FC system with batteries storage   14
Fig. 1.15   PV/FC system with battery storage block diagram   15
Fig. 1.16   PV/battery/FC system   15
Fig. 1.17   PV/FC system multi-storage batteries/super-capacitors   16
Fig. 1.18   PV/FC system with multi-storage batteries/super-capacitor
            block diagram   16
Fig. 1.19   Hybrid wind/photovoltaic system   17
Fig. 1.20   PV/wind system   18
Fig. 1.21   Aerogenerator model   18
Fig. 1.22   Wind turbine model   19
Fig. 1.23   Hybrid wind/photovoltaic system with battery storage   19
Fig. 1.24   Wind/photovoltaic system with battery storage   20
Fig. 1.25   Hybrid wind/photovoltaic system with flywheel storage   20
Fig. 1.26   Hybrid wind/photovoltaic system   21
Fig. 1.27   Hybrid wind/fuel cell system with battery storage   22
Fig. 1.28   Hybrid wind/fuel cell system with hybrid storage   22

Fig. 1.29   Hybrid photovoltaic system/diesel generators   23
Fig. 1.30   Hybrid wind/photovoltaic system with battery storage   23
Fig. 1.31   Hybrid wind/photovoltaic system   24
Fig. 1.32   Hybrid wind/photovoltaic system/diesel generators with
            battery storage   25
Fig. 1.33   Wind/photovoltaic system/diesel generators with battery
            storage under MATLAB/Simulink   25
Fig. 2.1   Various static converters depending on loads   40
Fig. 2.2   DC/DC converter in PV system   40
Fig. 2.3   DC/DC converter with MPPT controller in standalone
            PV system   41
Fig. 2.4   Buck converter . . . . :   41
Fig. 2.5   Configuration between time 0 and $D.T$ 42
Fig. 2.6   Configuration between time $D.T$ and T   42
Fig. 2.7   Boost converter   43
Fig. 2.8   First boost configuration   43
Fig. 2.9   Second boost configuration   44
Fig. 2.10   Simulink model of the boost converter   45
Fig. 2.11   Output voltage regulation in a boost converter   45
Fig. 2.12   Current ripple waveforms   46
Fig. 2.13   Voltage ripple   47
Fig. 2.14   Developed boost converter   48
Fig. 2.15   Measured boost voltage, current, power and control signal   48
Fig. 2.16   Buck–boost converter   49
Fig. 2.17   Non-inverting buck–boost converter   49
Fig. 2.18   Inverting buck–boost converter   49
Fig. 2.19   Example of DC/DC converters in PV system   50
Fig. 2.20   Example of DC/DC converter in wind system   51
Fig. 2.21   Examples of DC/DC converter in storage systems   52
Fig. 2.22   DC/DC converters in hybrid PV/wind system with battery
            storage   53
Fig. 2.23   DC/DC converters in PV with multi-storage FCs/batteries   54
Fig. 2.24   Buck–boost model under MATLAB/Simulink   54
Fig. 2.25   DC/AC converter   55
Fig. 2.26   General electrical schematic of three-phase inverter   55
Fig. 2.27   Inverter control   56
Fig. 2.28   DC/AC inverter model with PWM under
            MATLAB/Simulink   56
Fig. 2.29   Phase voltage and switch signals in DC/AC converter   57
Fig. 2.30   Hysteresis current control principle   57
Fig. 2.31   Determination of voltage with hysteresis control current   58
Fig. 2.32   Inverter current control   58
Fig. 2.33   Control hysteresis current under MATLAB/Simulink   59
Fig. 2.34   Simple DC system   59

Fig. 2.35   Simplified model of the battery charger   60
Fig. 2.36   Single-phase voltage inverter   61
Fig. 2.37   Three-phase voltage inverter.......................   61
Fig. 2.38   Example of double-stage inverter   62
Fig. 2.39   Dual-stage inverter   62
Fig. 2.40   Multi-input inverter......................   63
Fig. 2.41   Central plant inverter   65
Fig. 2.42   String inverter.......................   66
Fig. 2.43   AC modules   67
Fig. 2.44   Multi-string inverter   68
Fig. 2.45   Direct bus control.......................   69
Fig. 2.46   Calcul of current references   70
Fig. 2.47   Power control of PV system connected to the grid...........   71
Fig. 2.48   PV pumping system with FOC strategy   72
Fig. 2.49   PV pumping system with DTC strategy   73
Fig. 2.50   DC/DC converter in wind turbine system..................   73
Fig. 2.51   Wind system with controlled rectifier......................   73
Fig. 2.52   Uncontrolled rectifier   74
Fig. 2.53   The back-to-back PWM-VSI converter in wind system   74
Fig. 2.54   Connection of PV generator to the grid   74
Fig. 2.55   PV system using multi-cell inverter and controlled with
            FOC method........................   75
Fig. 2.56   Used converters in PV/DG/wind system with battery
            storage   75
Fig. 2.57   Used converters in PV/wind system   76
Fig. 3.1   Example of a hybrid PV/wind turbine system with MPPT
            controllers........................   80
Fig. 3.2   Photovoltaic electrical characteristics   81
Fig. 3.3   Variation of the voltage depending on the power............   82
Fig. 3.4   Flowchart of the P&O MPPT algorithm...................   82
Fig. 3.5   P&O algorithm under MATLAB/Simulink..................   83
Fig. 3.6   Maximum power point tracking code for a photovoltaic
            panel........................   84
Fig. 3.7   PV system with P&O MPPT   85
Fig. 3.8   Simulation results of P&O method in PV system............   86
Fig. 3.9   Modified P&O model.......................   86
Fig. 3.10   Modified P&O under MATLAB/Simulink   87
Fig. 3.11   Incremental conductance algorithm principle   88
Fig. 3.12   IncCond algorithm   88
Fig. 3.13   Flowchart of the Inc MPPT algorithm   89
Fig. 3.14   IncCond MPPT method under MATLAB/Simulink   89
Fig. 3.15   IncCond MPPT simulation results   90
Fig. 3.16   Characteristic power-voltage $P_{pv}(V_{pv})$ of PV generator........   91
Fig. 3.17   Flowchart of the OCV MPPT algorithm...................   91

Fig. 3.18   Flowchart of the SC MPPT algorithm 92
Fig. 3.19   Block diagram of temperature method 93
Fig. 3.20   PSO concept.. 95
Fig. 3.21   PSO algorithm 95
Fig. 3.22   PSO flowchart....................................... 96
Fig. 3.23   Voltage particles motion.............................. 97
Fig. 3.24   Current particles motion 97
Fig. 3.25   Objective function of particles 98
Fig. 3.26   Ant colony algorithm 98
Fig. 3.27   Flowchart of the ACO algorithm 99
Fig. 3.28   Membership functions for: **a** input variable E, **b** input variable
            CE, **c** output variable D 100
Fig. 3.29   FLC MPPT under MATLAB/Simulink.................... 100
Fig. 3.30   Simulation results with FLC........................... 101
Fig. 3.31   Membership functions of AFLC method.................. 102
Fig. 3.32   AFLC model under MATLAB/Simulink.................. 103
Fig. 3.33   Simulation results with AFLC 103
Fig. 3.34   Sliding mode control algorithm under MATLAB/Simulink 105
Fig. 3.35   Simulation results using SMC.......................... 105
Fig. 3.36   Example of a neural network 106
Fig. 3.37   Photovoltaic system with MPPT control by neuro-fuzzy
            network.. 106
Fig. 3.38   Block diagram of NF MPPT............................ 107
Fig. 3.39   Block diagram of a PV system based on genetic algorithm 108
Fig. 3.40   HP&OIncCond MPPT under MATLAB/Simulink 108
Fig. 3.41   Simulation results with hybrid P&O/IncCond method 109
Fig. 3.42   Block diagram of improved MPPT using FLC.............. 110
Fig. 3.43   Comparison of most important MPPT methods 116
Fig. 3.44 $P_{pv}(V_{pv})$ characteristic with partial shading.............. 118
Fig. 3.45   Partial shading in MATLAB/Simulink 118
Fig. 3.46 $I_{pv}(V_{pv})°$ and $P_{pv}(V_{pv})$ characteristics with partial shading 119
Fig. 3.47   Reference torque as a function of speed 120
Fig. 3.48   Tip speed ratio method of wind energy conversion system..... 121
Fig. 3.49   Block diagram of WTCS with TSR MPPT method 121
Fig. 3.50   Power signal feedback control of wind energy conversion
            system... 122
Fig. 3.51   Block diagram of WTCS with PSF method 122
Fig. 3.52   PSF under MATLAB/Simulink.......................... 122
Fig. 3.53   Optimal torque control of wind energy conversion system..... 124
Fig. 3.54   OTC MPPT under MATLAB/Simulink 124
Fig. 3.55   HCS principle.. 125
Fig. 3.56   HCS MPPT under MATLAB/Simulink 125
Fig. 3.57   Gradient method principle.............................. 126
Fig. 3.58   Flowchart of GM 127

Fig. 3.59   Gradient method under MATLAB/Simulink.   128
Fig. 3.60   Hybrid HCS/OTC MPPT in wind turbine system.   128
Fig. 3.61   Application of HP&O/OTC in stand-alone wind turbine system . . .   128
Fig. 3.62   Membership of the error (*E*), the error change (CE)
            and the output ω_{ref}   129
Fig. 3.63   ANN control of wind energy conversion system   130
Fig. 3.64   Wind energy conversion system with RBFN controller   131
Fig. 3.65   General structure of neuro-fuzzy controller.   131
Fig. 3.66   Comparison of most used MPPT in wind turbine system.   133
Fig. 4.1   Types of energy storage systems   140
Fig. 4.2   Most used storage technologies   140
Fig. 4.3   Battery storage with buck–boost converter.   142
Fig. 4.4   Batteries storage in hybrid system   143
Fig. 4.5   PV system with batteries supplying a load   148
Fig. 4.6   PV system with battery storage.   148
Fig. 4.7   Battery model in MATLAB/Simulink.   149
Fig. 4.8   Simulation results of a battery under MATLAB/Simulink   150
Fig. 4.9   FEES with IG in wind turbine system   151
Fig. 4.10   WCES based on an induction generator with flywheel
            storage   151
Fig. 4.11   Flywheel storage model under MATLAB/Simulink   152
Fig. 4.12   Simulation results using flywheel in wind turbine system.   153
Fig. 4.13   SCs with buck–boost converter.   154
Fig. 4.14   SCs model under MATLAB/Simulink   155
Fig. 4.15   Equivalent electrical circuit diagram of a PEMFC   156
Fig. 4.16   Block diagram of PEMFCs model   156
Fig. 4.17   Fuel cells voltage and power   157
Fig. 4.18   Fuel cells connected to boost converter   157
Fig. 4.19   Equivalent electrical circuit of SCs.   158
Fig. 4.20   SCs model under MATLAB/Simulink   159
Fig. 4.21   SC voltage, current and state of charge   160
Fig. 4.22   Hybrid FC/SC storage in MATLAB/Simulink   161
Fig. 4.23   Simulation results of hybrid FC/CS storage   162
Fig. 4.24   Structure of the hybrid storage SCs/FCs supplying EV system.   162
Fig. 4.25   Hybrid storage SCs/FCs supplying DC motor   163
Fig. 4.26   Hybrid storage battery/FCs supplying DC motor   163
Fig. 4.27   Fuel cell/batteries system for EVs.   164
Fig. 4.28   Hybrid FC/battery storage under MATLAB/Simulink.   164
Fig. 4.29   Hybrid batteries/FC storage in wind turbine system   165
Fig. 4.30   Algorithm of wind turbine with batteries and fuel cells   166
Fig. 4.31   Hybrid battery/SC storage supplying DC motor   167
Fig. 4.32   Super-capacitor/battery system   167

Fig. 4.33   Hybrid storage in wind turbine system   168
Fig. 4.34   FC/SC/battery storage system   169
Fig. 5.1   Example of a time schedule diagram   174
Fig. 5.2   Obtained results of PV/wind design. **a** Solar radiation and
            wind speeds of the location. **b** Average daily energy of PV
            and wind generators. **c** Full energy with PV and wind turbine
            energies. **d** Combinations of number of PV
            and wind turbines   183
Fig. 5.3   Diagram of a typical standalone PV system powering
            AC loads   184
Fig. 5.4   Variation of powers   187
Fig. 5.5   Example hybrid combinations systems in Homer Pro.
            a PV/batteries. **b** PV/wind/batteries. **c** PV/wind/batteries/HPS.
            d PV/wind/HPS/FEES. **e** PV/wind/batteries on grid.
            f PV/wind/batteries/batteries on grid   188
Fig. 5.6   Bejaia location in Algeria (Latitude 36°45.3522′ N, Longitude
            5°5.0598′ E) with HOMER software License Agreement   189
Fig. 5.7   Daily radiation and clearness index at Bejaia location
            (downloaded at 18/08/2019 18:58:26 from HOMER software
            License Agreement)   189
Fig. 5.8   Average wind speeds (downloaded at 18/08/2019 18:58:26
            from HOMER software License Agreement)   190
Fig. 5.9   Daily average load profile for a residential house   190
Fig. 5.10   Scaled data monthly average.   191
Fig. 5.11   Daily average load for a complete year   191
Fig. 5.12   Hourly average load variations in a year for all months   192
Fig. 5.13   Block diagram of PV/wind turbine/battery hybrid system.   192
Fig. 6.1   Key decision factors for PMC   198
Fig. 6.2   Photovoltaic/battery system.   198
Fig. 6.3   Photovoltaic/battery system with PMC (first structure)   199
Fig. 6.4   Flowchart of PV/battery system-first configuration   200
Fig. 6.5   Power flow of PV/batteries system (first configuration)   201
Fig. 6.6   Chosen load profile.   202
Fig. 6.7   Temperature profile during four different days   202
Fig. 6.8   Solar irradiance profile during four different days.   203
Fig. 6.9   Evolution of the different powers during four different days. . . .   203
Fig. 6.10   Comparison between the sum of the different source powers
            and the load power   203
Fig. 6.11   Power flow of PV/battery system (second structure).   204
Fig. 6.12   Energy management strategy of PV/battery system based
            on four switches   205
Fig. 6.13   Schematic power flow of PV/batteries system
            (second configuration).   206
Fig. 6.14   Power management of PV/FC system.   208

Fig. 6.15   Flowchart of the PV/FC system 208
Fig. 6.16   Different power waveforms............................. 209
Fig. 6.17   Sum of the different source powers, the load power
            and the different modes............................. 209
Fig. 6.18   Photovoltaic/battery/fuel cell system..................... 210
Fig. 6.19   Flowchart of the third structure........................ 211
Fig. 6.20   Different modes of operation of the hybrid system studied..... 212
Fig. 6.21   Power variations of the different sources 213
Fig. 6.22   Sum of the different source powers and the load power 213
Fig. 6.23   Different modes under MATLAB/Simulink 214
Fig. 6.24   Overall block diagram of the studied system 214
Fig. 6.25   Simulink model for PMC of wind turbine/battery
            configuration...................................... 216
Fig. 6.26   Supervision model 216
Fig. 6.27   Wind speed profile 216
Fig. 6.28   Different switches K_1, K_2, K_3 and K_4 217
Fig. 6.29   Power waveforms during the two different days............. 217
Fig. 6.30   Proposed PV/wind/batteries for standalone system 218
Fig. 6.31   Power management algorithm of PV/wind/batteries 219
Fig. 6.32   PV/wind/batteries for standalone system................... 220
Fig. 6.33   Proposed PV/wind/batteries for pumping water system........ 220
Fig. 6.34   Control of DC bus with FOC in hybrid pumping system 221
Fig. 6.35   Simulink block diagram of PV/wind/battery system supplying
            a water pump 222
Fig. 6.36   Solar radiation and wind speeds 223
Fig. 6.37   Different modes of operation 224
Fig. 6.38   Hybrid, battery and load power 224
Fig. 6.39   Diagram of fuzzy controller for hybrid
            photovoltaic/wind/diesel system 225
Fig. 6.40   Flowchart of PMC for hybrid photovoltaic/wind/diesel system ... 227
Fig. 6.41   Block system of the hybrid PV/wind/diesel generator system
            with battery storage 228
Fig. 6.42   Profile of the solar irradiation and wind speeds 228
Fig. 6.43   Control signals of the three generators (PV, wind, diesel) 229
Fig. 6.44   Power waveforms 230
Fig. 6.45   PV/wind turbine/FC/battery hybrid structure with PMC 231
Fig. 6.46   Flowchart of hybrid system with battery storage 232
Fig. 6.47   Simulink block diagram of PV/wind/battery system supplying
            a water pump 233
Fig. 6.48   Solar irradiance and temperature profile 233
Fig. 6.49   Wind turbine profile 234
Fig. 6.50   Power variations of the different sources 234
Fig. 6.51   PV system with hybrid storage (batteries/super-capacitors)..... 235

Fig. 6.52   Control of the DC/DC boost converter of battery   235
Fig. 6.53   Control of the DC/DC boost of super-capacitors   236
Fig. 6.54   Control of PV system with hybrid storage
            (batteries/super-capacitors)   236
Fig. 6.55   PV supplying electric vehicle with hybrid storage   237
Fig. 6.56   Different forces acting on a vehicle moving along a slope   237
Fig. 6.57   PV/FCS with hybrid storage (batteries/SCs)   239
Fig. 6.58   Flowchart of the supervision of hybrid storage associated
            to PV source   240
Fig. 6.59   Supervision of PV/FCs with hybrid storage (batteries/SCs)   240
Fig. 6.60   Supervision of FC/battery system supplying electric vehicle:
            first structure   241
Fig. 6.61   Different modes of the electric vehicle   242
Fig. 6.62   Flowchart of power management of fuel cells/batteries system
            supplying electric vehicle (first structure)   243
Fig. 6.63   Supervision of FC/battery system supplying electric vehicle:
            second structure   245
Fig. 6.64   Flowchart of power management of FC/battery system
            supplying electric vehicle (second structure)   245

List of Tables

Table 1.1   Classification of hybrid systems by power range   6
Table 1.2   Different hybrid systems   7
Table 1.3   Different alternatives with only two components   7
Table 1.4   Different alternatives considering multiple storages   8
Table 1.5   Summary of the most used hybrid systems   9
Table 2.1   Boost converter experimental results under different tests   48
Table 3.1   Rule table of modified P&O   87
Table 3.2   Fuzzy rules table   99
Table 3.3   Modified fuzzy rules table   101
Table 3.4   Classical and advanced P&O algorithms   111
Table 3.5   Classical and modified algorithms IncCond   112
Table 3.6   Other MPPT methods based on only or without sensor   113
Table 3.7   MPPT based on two sensors   114
Table 3.8   Advanced MPPT algorithms   115
Table 3.9   Possible cases for GM   126
Table 3.10 Fuzzy rules used for the fuzzy controller   129
Table 3.11 Some important proprieties of the most used MPPT
            methods   134
Table 4.1   Some important parameters of battery, FCs and SC [2–4]   141
Table 4.2   Most battery types   141
Table 4.3   Battery models based on equations   143
Table 4.4   An overview of simple battery models based on circuits   144
Table 4.5   An overview of battery Thevenin models   144
Table 4.6   Linear and nonlinear battery models   145
Table 4.7   Dynamic battery models   145
Table 4.8   Other used battery models   146
Table 5.1   Results of Application 2   174
Table 5.2   Results of the Application 3   176
Table 5.3   Results of the Application 4   177
Table 5.4   Results of the application 5   178
Table 5.5   Results of the Application 6   179

Table 5.6   Monthly energies produced by photovoltaic and wind
               generators.   181
Table 5.7   Sizing according to the annual monthly average   182
Table 5.8   Results of Application 9.   184
Table 5.9   Results of application 8   186
Table 5.10   Moto-pump group sizing   186
Table 6.1   Status of the switches and the different modes.   201
Table 6.2   Simplified table based on four switches of PV/battery
               system   206
Table 6.3   Switches state operating modes in the second structure   209
Table 6.4   Switches state operating modes in PV/batteries/FCs.   211
Table 6.5   Different modes in wind turbine/battery system   215
Table 6.6   Fuzzy inference of fuzzy controller inputs/outputs   225
Table 6.7   Different values of E_s, V_{wind} and SOC.   226
Table 6.8   Operation modes for hybrid photovoltaic/wind/diesel
               generator system   227
Table 6.9   Modes of vehicle operation   242
Table 6.10   Simplified table in the case of three switches.   244
Table 6.11   Table in the second structure   246

Chapter 1
Hybrid Renewable Energy Systems Overview

1.1 Introduction

Wind and photovoltaic sources are one of the cleaner forms of energy conversion available. One of the advantages offered by the hybridization of different sources is to provide sustainable electricity in areas not served by the conventional power grid. They are very used in many applications, but due to their nonlinearity, hybrid energy systems are proposed to overcome this problem with important improvements [1–204]. In general, hybridization consists of combining several energy sources and storage units within the same system in order to optimize the production and energy management. In review papers, they can be found under the following names: hybrid renewable energy systems (HRESs) or multi-source multi-storage systems (MSMSSs).

1.2 Advantages and Disadvantages of an Hybrid System

Hybrid renewable energy systems (HRESs) are attractive configurations used for different applications and especially in standalone power generation systems as electrification, water pumping and telecommunications. The most advantages of these systems are their simplicity to use and their independent from one energy source, so they can be productive during the day the night. On the other side, the disadvantage is that there are different sources and storage units, so the system is more complex than a single-source system. In this case, an energy management control is necessary to control the power flow, so the global system will be more complex and of course higher cost [42, 62].

© Springer Nature Switzerland AG 2020
D. Rekioua, *Hybrid Renewable Energy Systems*, Green Energy and Technology,
https://doi.org/10.1007/978-3-030-34021-6_1

1.3   Configuration of Hybrid System

The first and most basic decision that a power system designer is faced is what architecture to be used. This decision will influence every other aspects of the system design including the types and quantities of power converters that will be needed. So two choices must be considered [42, 54]:

- Choice of power converters
- Choice of common bus type.

1.3.1   Choice of Common Bus Type

The different energy sources can been interconnected through a DC bus or through an AC bus or through DC/AC bus [43, 55–62].

1.3.1.1   Architecture of DC Bus

In the hybrid system presented in Fig. 1.1, the power supplied by each source is centralized on a DC bus. Thus, the energy conversion system to provide AC power

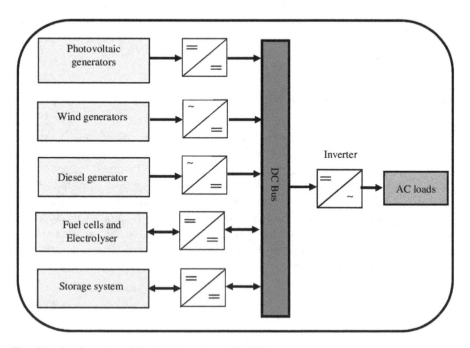

Fig. 1.1 Configuration of the hybrid system with DC bus

at their first rectifier has to be converted then continuously. The generators are connected in series with the inverter to power the load alternatives. The inverter should supply the alternating loads from the DC bus and must follow the set point for the amplitude and frequency. The batteries are sized to supply peak loads. The advantage of this topology is the simplicity of operation, and the load demand is satisfied without interruption even when the generators charge the short-term storage units.

1.3.1.2 Architecture of AC Bus

In this topology, all components of the HPS are related to alternating loads, as shown in Fig. 1.2. This configuration provides superior performance compared to the previous configuration, since each converter can be synchronized with the generator so that it can supply the load independently and simultaneously with other converters. This provides flexibility for the energy sources which supply the load demand. In the case of low load demand, all generators and storage systems are stationary except, for example, the photovoltaic generator to cover the load demand. However, during heavy load demands or during peak hours, generators and storage units operate in parallel to cover the load demand. The realization of this system is relatively complicated because of parallel operation, by

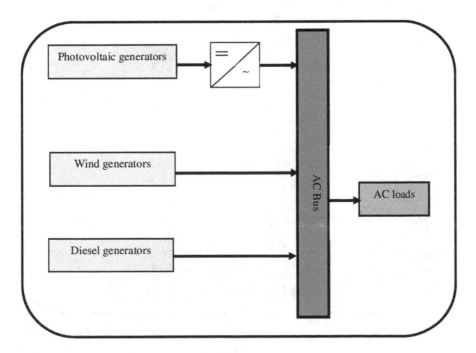

Fig. 1.2 Configuration of the hybrid system with AC bus

synchronizing the output voltages with the charge voltages. This topology has several advantages compared to the DC-coupled topology such as higher overall efficiency, smaller sizes of the power conditioning unit while keeping a high level of energy availability, and optimal operation of the diesel generator due to reducing its operating time and consequently its maintenance cost.

1.3.1.3 Architecture of DC/AC Bus

The configuration of DC and AC buses is shown in Fig. 1.3. It has superior performance compared to the previous configurations. In this case, renewable energy and diesel generators can power a portion of the load directly to AC, which can increase system performance and reduce power rating of the diesel generator and the inverter. The diesel generator and the inverter can operate independently or in parallel by synchronizing their output voltages. Converters located between two buses (the rectifier and inverter) can be replaced by a bidirectional converter which, in normal operation, performs the conversion DC/AC (inverter operation). When there is a surplus of energy from the diesel generator, it can also charge batteries (operating as a rectifier). The bidirectional inverter can supply the peak load when the diesel generator is overloaded.

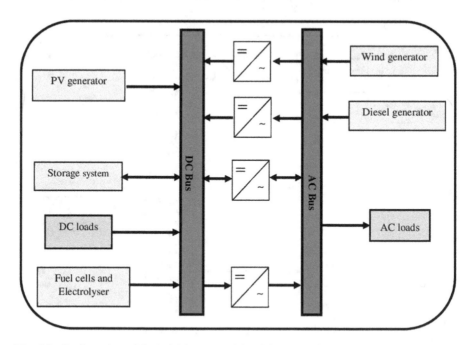

Fig. 1.3 Configuration of the hybrid system with AC bus and DC bus

The advantages of this configuration are:

- The diesel generator and the inverter can operate independently or in parallel. When the load level is low, one or the other can generate the necessary energy. However, both sources can operate in parallel during peak load.
- The possibility of reducing the nominal power of the diesel generator and the inverter without affecting the system's ability to supply peak loads.

The disadvantages of this configuration are:

- The implementation of this system is relatively complicated because of the parallel operation (the inverter should be able to operate autonomously and operate with synchronization of the output voltages with output voltages of diesel generator).

1.3.2 Choice of Converters

A power converter is a system for adapting the source of electrical energy to a given receiver by converting it (Fig. 1.4).

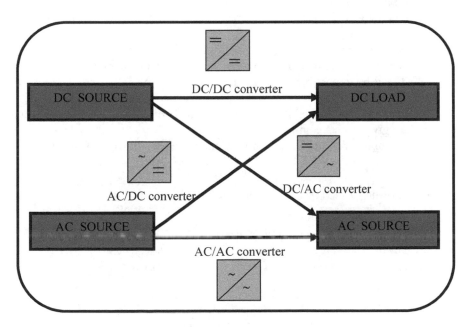

Fig. 1.4 Sources and loads supplied by various static converters

1.4   Classifications of Hybrid Energy Systems

The power delivered by the hybrid system can vary from a few watts for domestic applications up to a few megawatts for systems used in the electrification of small islands. Thus, for hybrid systems with a power below 100 kW, the configuration with AC and DC bus, with battery storage, is the most used. The storage system uses a high number of batteries to be able to cover the average load for several days. This type of hybrid system uses small renewable energy sources connected to the DC bus. Another possibility is to convert the continuous power to an alternative one by using inverters. Hybrid systems used for applications with very low power (below 5 kW) supply generally DC loads (Table 1.1).

1.5   Different Combinations of Hybrid Systems

Mathematically, it can have 2 power n (2^n) combinations of hybrid systems. In the following, the most used combinations of hybrid system are presented as follows (Fig. 1.5).

Mathematically, it can have the following combinations with one storage (Tables 1.2 and 1.3).

By combining just one element with another, there are about eighteen alternatives.

And by considering multiple storages, it can obtain a multiplicity of configurations (about seventy). Some of them have been cited in the literature, others not at all, which may be impossible to do because of the complexity of some combinations (Table 1.4).

The most important systems are presented, and references of the most cited systems are given to have an overview.

The most used hybrid systems can be summarized as shown in Table 1.5.

Table 1.1 Classification of hybrid systems by power range

Hybrid system power	Applications
Low power	Autonomous systems: pumping water, telecommunication stations, ...
Average power	Micro-isolated systems: supplying village, rural...
Great power	Large isolated systems, for example, islands

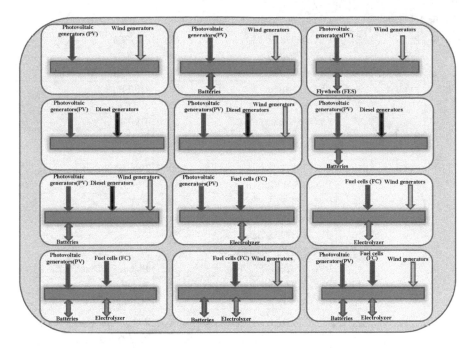

Fig. 1.5 Representation of some used hybrid systems

Table 1.2 Different hybrid systems

	PVG	Batt	FCs	SCs	WT	FES	DG	HP
PVG								
Batt	x							
FCs	x	**x**						
SCs	x	**x**	x					
WT	x	x	x	x				
FES	x	**x**	**x**	**x**	x			
DG	x	**x**	**x**	**x**	x			
HP	x	x	x	x	x	x	x	

Table 1.3 Different alternatives with only two components

1	PVG/Batt	10	WT/SCs
2	PVG/FCs	11	WT/FES
3	PVG/SCs	12	HP/FCs
4	PVG/WT	13	HP/SCs
5	PVG/DG	14	WT/HP
6	PVG/HP	15	HP/FES
7	WT/Batt	16	WT/DG
8	HP/Batt	17	HP/DG
9	WT/FCs	18	PV/FES

Table 1.4 Different alternatives considering multiple storages

1	PVG/Batt	2	PVG/WT/Batt	43	HP/FCs
		3	PVG/WT/SCs	44	HP/SCs
		4	PVG/WT/Batt/SCs	45	WT/HP
5	PVG/FCs	6	PVG/FCs/Batt	46	HP/FES
		7	PVG/FCs/SCs	47	WT/DG
		8	PVG/FCs/Batt/SCs	48	HP/DG
9	PVG/SCs	10	PVG/SCs/Batt	49	PVG/FES
11	PVG/WT	12	PVG/WT/FES	50	PVG/FES/Batt
		13	PVG/WT/FES/Batt	51	PVG/FES/FCs
		14	PVG/WT/FES/SCs	52	PVG/FES/SCs
		15	PVG/WT/FES/FCs	53	PVG/FES/Batt/FCs
		16	PVG/WT/FES/Batt/FCs	54	PVG/FES/Batt/SCs
		17	PVG/WT/FES/Batt/SCs	55	PVG/FES/FCs/SCs
		18	PVG/WT/FES/FCs/SCs	56	PVG/FES/Batt/FCs/SCs
		19	PVG/WT/FES/Batt/FCs/SCs	57	WT/Batt
		20	PVG/WT/DG	58	HP/Batt
		21	PVG/WT/DG/Batt	59	WT/FCs
		22	PVG/WT/DG/FCs	60	WT/SCs
		23	PVG/WT/DG/SCs	61	WT/FES
		24	PVG/WT/DG/Batt/FCs	62	WT/FES/SCs
		25	PVG/WT/DG/Batt/SCs	63	WT/FES/FCS
		26	PVG/WT/DG/FCs/SCs	64	WT/FES/Batt
		27	PVG/WT/DG/Batt/FCs/SCs	65	WT/FES/SCs/Batt
		28	PVG/WT/DG/FES	66	WT/FES/SCs/FCs
		29	PVG/WT/DG/FES/Batt	67	WT/FES/FCs/Batt
		30	PVG/WT/DG/FES/FCs	68	WT/FES/SCs/Batt/FCs
		31	PVG/WT/DG/FES/SCs	69	PVG/HP
		32	PVG/WT/DG/FES/Batt/FCs	70	PVG/HP/Batt
		33	PVG/WT/DG/FES/Batt/SCs	71	PVG/HP/Batt/FCs
		34	PVG/WT/DG/FES/FCs/SCs	72	PVG/HP/Batt/FCs/WT
		35	PVG/WT/DG/FES/Batt/FCs/SCs	73	PVG/HP/Batt/FCs/WT/FES
5	PVG/DG	36	PVG/DG/Batt	74	PVG/HP/Batt/FCs/WT/FES/DG
		37	PVG/DG/FCs		
		38	PVG/DG/SCS		
		39	PVG/DG/Batt/FCs		
		40	PVG/DG/Batt/SCs		
		41	PVG/DG/FCs/SCs		
		42	PVG/DG/Batt/FCs/SCs		

Table 1.5 Summary of the most used hybrid systems

Hybrid systems	Some references
PVG/Batt	[14, 15, 42, 76–88]
PVG/FCs	[22, 54, 62, 10, 116, 117, 145]
PVG/FCs/Batt	[30, 55, 52, 184, 185]
PVG/Wind	[47, 65, 112, 113, 142]
PVG/Wind/Batt	[123–127]
PVG/Wind/FES	[193–195]
WT/FCs	[1, 5, 23, 171, 178]
WT/FCs/Batt/SCs	[196–198]
WT/FCs/Batt	[182, 183]
PVG/FCs/DG	[24, 26, 31, 33, 39, 51, 26, 52, 63, 51]
PVG/DG	[19, 28, 32, 34, 37, 38, 40, 94, 95]
PVG/Batt/DG	[4, 12, 25, 29, 41, 98–100]
PVG/WT/DG	[96, 97, 128, 129]
PVG/WT/FCs/Batt	[120, 188]
PVG/WT/FCs/DG	[186, 190]
PVG/DG/FES	[189, 199]
PVG/Batt/DG/FCs	[2, 11, 2]
PVG/WT/Batt/DG	[13, 13, 102, 111, 151]
PVG/WT/FCs	[118, 119, 121, 122, 146]
PVG/HP/Batt	[104, 130–132, 130]
PVG/Batt/SCs	[103, 133–137]
PVG/WT/FCs	[3, 6, 10, 16–18, 44, 48, 53, 59, 64]
PVG/HP	[105, 106, 114, 179, 180]
PVG/DG/HP	[27, 98, 147, 148]
PVG/DG/HP/FCs	[187, 191]
WT/DG/Batt	[140, 142, 170]
WT/HP	[100, 101, 109, 181]
WT/Batt	[68, 154, 162, 168, 174]
WT/HP/DG	[192, 200, 201]
WT/PVG/HP	[104, 107, 108, 110, 115, 149]
WT/DG	[141, 153–158, 164, 176]
WT/PVG/HP/DG	[147, 150]
WT/SCs	[159, 169, 172, 173]
WT/FES	[160, 161, 166, 167, 177]
WT/DG/FES	[163, 165, 175]
PVG/WT/FCs/DG/UC	[202]
PVG/WT/SCs/Batt	[203, 204]
PVG/FCs/Batt/SCs	[90, 93, 103]

1.5.1 PV System with Battery Storage

In standalone PV applications, electrical power is required from the system during night or hours of darkness [14, 15]. Thus, the storage must be added to the system. Generally, batteries are used for energy storage (Fig. 1.6).

This system can supply DC and AC loads (Fig. 1.7) [42, 49–88].

It can be implemented under MATLAB/Simulink as shown in Fig. 1.8.

And the different subsystems are in Fig. 1.9.

1.5.2 PV System/Fuel Cells

The role of PV/FCs system is the production of electricity without interruption in remote areas. It consists generally of a photovoltaic generator (PV), an alkaline water electrolyzer, a storage gas tank and a proton exchange membrane fuel cell (PEMFC) (Fig. 1.10) [22, 54].

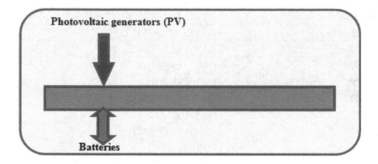

Fig. 1.6 Photovoltaic system with battery storage

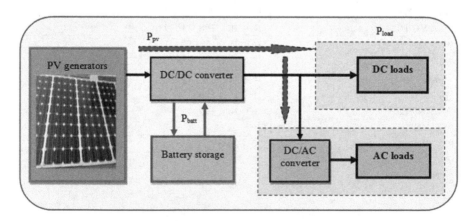

Fig. 1.7 Standalone PV system with battery storage powering DC and AC loads

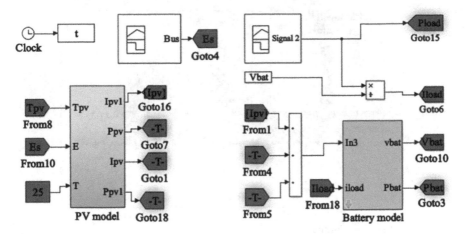

Fig. 1.8 PV/batteries under MATLAB/Simulink

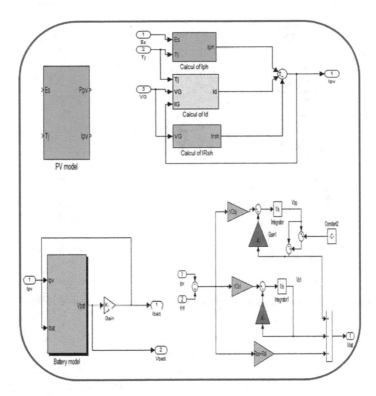

Fig. 1.9 Different blocks of PV/battery system

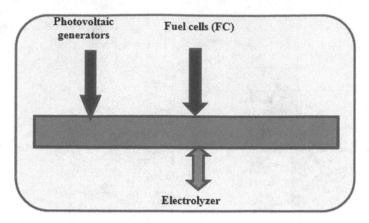

Fig. 1.10 PV system with fuel cells

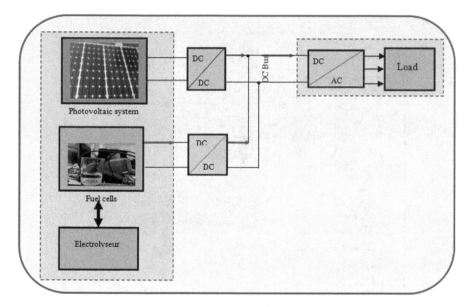

Fig. 1.11 Hybrid photovoltaic/fuel cell block diagram

PV subsystem works as a primary source, converting solar irradiation into electricity that is given to a DC bus (Fig. 1.11). The second working subsystem is the electrolyzer which produces hydrogen and oxygen from water as a result of an electrochemical process. When there is an excess of solar generation available, the electrolyzer is turned on to begin producing hydrogen which is sent to a storage tank. The produced hydrogen is used by the third working subsystem (the fuel cell stack) which produces electrical energy to supply the DC bus [62, 10, 116, 117, 145].

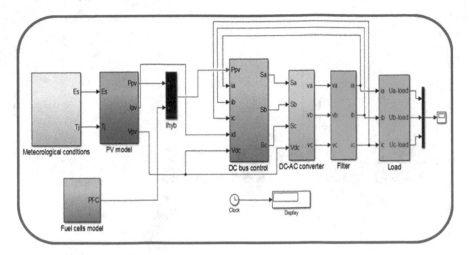

Fig. 1.12 PV/FC system

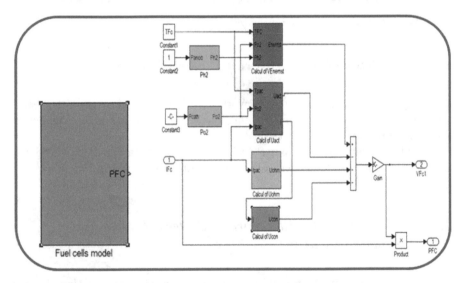

Fig. 1.13 Fuel cell model

It can be implemented under MATLAB/Simulink as shown in Fig. 1.12.
The fuel cell model is shown in Fig. 1.13.

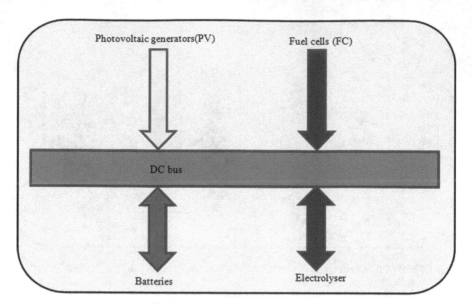

Fig. 1.14 PV/FC system with batteries storage

1.5.3 PV System/Fuel Cells with Battery Storage

In this system, PV subsystem always works as a primary source; then, the second working subsystems are the battery storage and the electrolyzer which supplies the fuel cells (Fig. 1.14) [30, 153–184].

The block diagram representing PV/FC system is given in Fig. 1.15 [43, 55–62].

It can be implemented under MATLAB/Simulink as shown in Fig. 1.16.

1.5.4 PV System/FC Multi-storage Batteries/ Super-Capacitors

In this system, it is added a multi-storage to the previous system (Fig. 1.15). It consists of batteries and super-capacitors (Fig. 1.17).

The block diagram representing PV/FC system with multi-storage is given in Fig. 1.18.

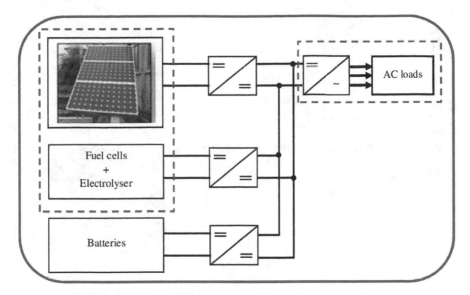

Fig. 1.15 PV/FC system with battery storage block diagram

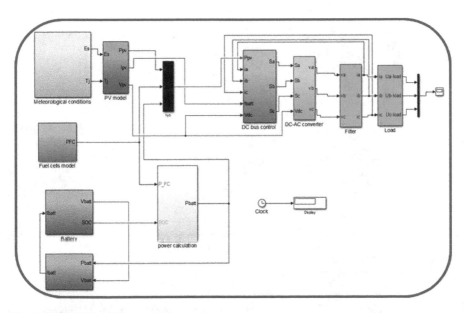

Fig. 1.16 PV/battery/FC system

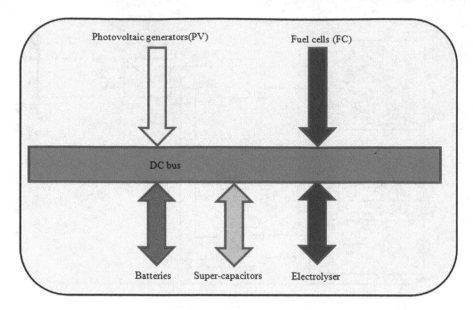

Fig. 1.17 PV/FC system multi-storage batteries/super-capacitors

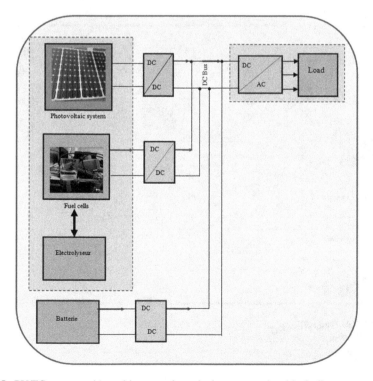

Fig. 1.18 PV/FC system with multi-storage batteries/super-capacitor block diagram

1.5.5 Hybrid Wind/Photovoltaic System

The advantage of this type of hybrid system depends on the wind, solar radiation and the type of load. It consists of a photovoltaic subsystem, a DC/DC converter and a wind turbine. The two energy sources are connected to a DC bus [47, 48, 112, 113, 143] (Fig. 1.19).

It can be implemented under MATLAB/Simulink as shown in Fig. 1.20.
The aerogenerator subsystem is modeled as shown in Fig. 1.21.
And the wind turbine model is shown in Fig. 1.22.

1.5.6 Hybrid Wind/Photovoltaic System with Battery Storage

Both energy sources are connected to a DC bus, and batteries are added as a storage system [123–127] (Fig. 1.23).

It can be implemented under MATLAB/Simulink as shown in Fig. 1.24.

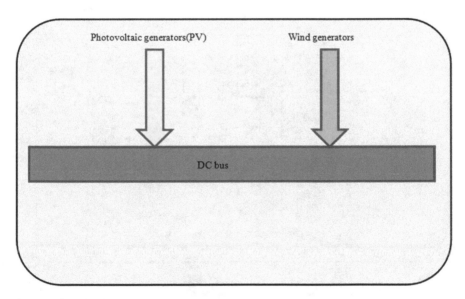

Fig. 1.19 Hybrid wind/photovoltaic system

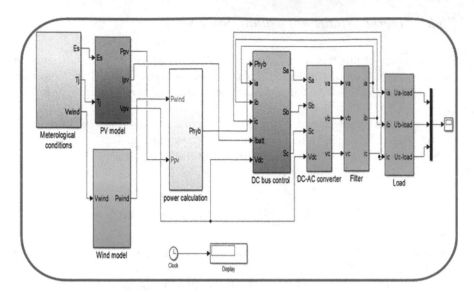

Fig. 1.20 PV/wind system

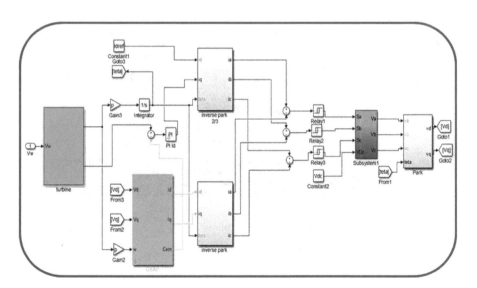

Fig. 1.21 Aerogenerator model

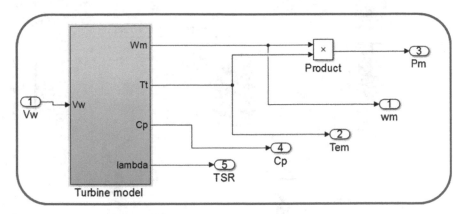

Fig. 1.22 Wind turbine model

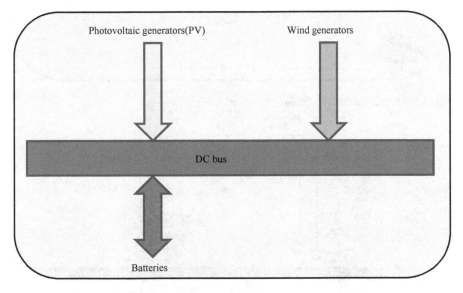

Fig. 1.23 Hybrid wind/photovoltaic system with battery storage

1.5.7 Hybrid Wind/Photovoltaic System with Flywheels Storage

Flywheels energy storage can also be used. FES works by accelerating a rotor (flywheel) to a very high speed and maintaining the energy in the system as rotational energy (Fig. 1.25) [193–195].

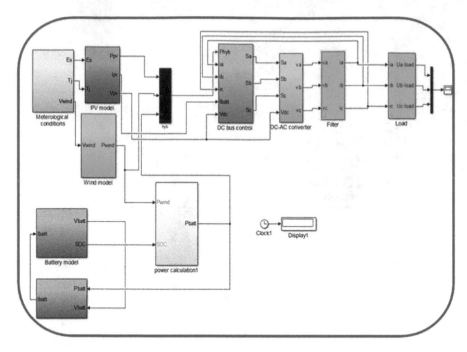

Fig. 1.24 Wind/photovoltaic system with battery storage

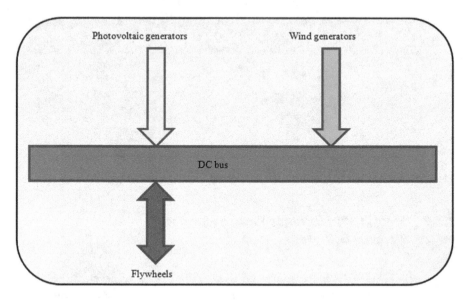

Fig. 1.25 Hybrid wind/photovoltaic system with flywheel storage

1.5.8 Wind Turbine System with Fuel Cells

The system consists of a wind generation system with an electrolyzer to generate hydrogen from surplus wind and a fuel cell for storage. Wind generator turbine provides electricity for electrolyzer, and the excess of energy can be send to generate hydrogen for storage and converted into electricity during peak times [1, 5, 23, 171–178] (Fig. 1.26).

1.5.9 Wind System/Fuel Cells with Battery Storage

Battery storage can be added to the previous system (Fig. 1.27) [182, 183].

1.5.10 Wind System/Fuel Cells with Hybrid Storage Batteries/Super-Capacitors

In this system, it is added a multi-storage. It consists of batteries and super-capacitors (Fig. 1.28) [196–198].

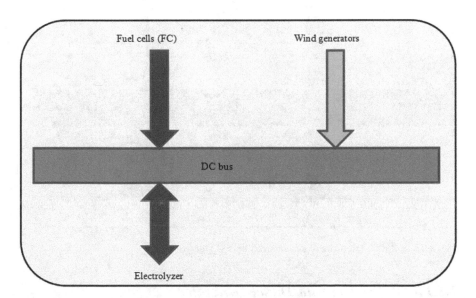

Fig. 1.26 Hybrid wind/photovoltaic system

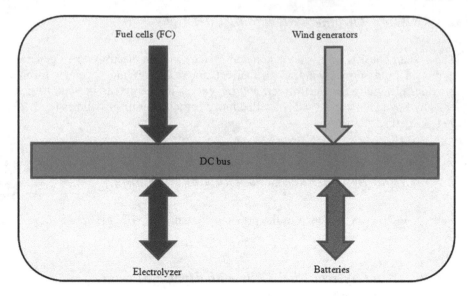

Fig. 1.27 Hybrid wind/fuel cell system with battery storage

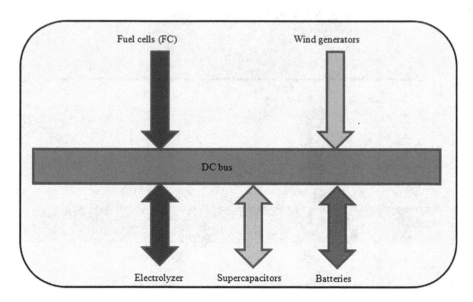

Fig. 1.28 Hybrid wind/fuel cell system with hybrid storage

1.5.11 PV System with Diesel Generators

It is the most used hybrid system. It comprises a photovoltaic generator with a diesel generator (Fig. 1.29) [19, 28, 32, 34, 37–40, 94, 95].

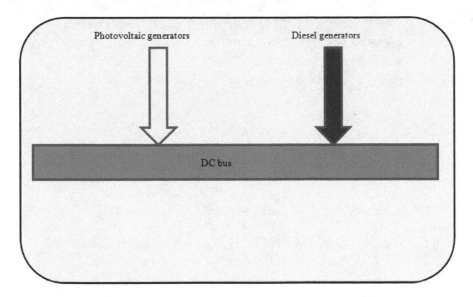

Fig. 1.29 Hybrid photovoltaic system/diesel generators

1.5.12 PV System with Diesel Generators with Battery Storage

Battery storage can be added to the previous system (Fig. 1.30) [4, 12, 25, 29, 41, 19, 98–100].

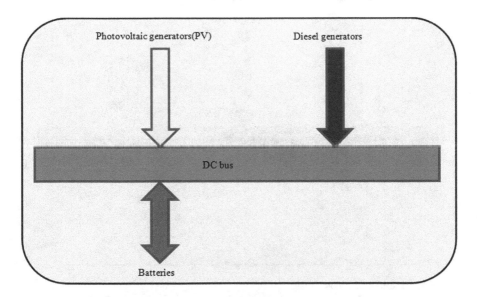

Fig. 1.30 Hybrid wind/photovoltaic system with battery storage

1.5.13 PV System with Wind Turbine System and Diesel Generators

Diesel generators are added as a backup system to PV/wind turbine system (Fig. 1.31).

1.5.14 PV System with Wind Turbine and Diesel Generators with Battery Storage

In this case, battery storage is added to the previous system (Fig. 1.32) [96, 97, 128, 129].

It can be implemented under MATLAB/Simulink as shown in Fig. 1.33.

1.6 Conclusion

This chapter has been devoted to hybrid wind systems. The different configurations and the different combinations of hybrid wind systems have been presented and described. Different synoptic schemes and models are also presented to show their implementation under MATLAB/Simulink.

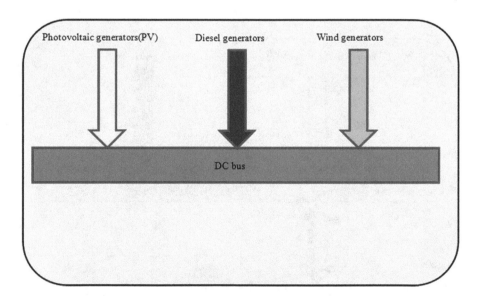

Fig. 1.31 Hybrid wind/photovoltaic system

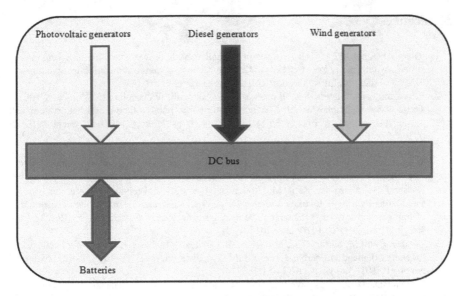

Fig. 1.32 Hybrid wind/photovoltaic system/diesel generators with battery storage

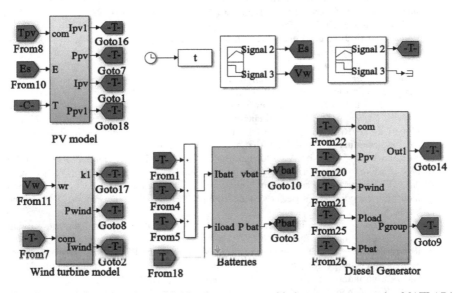

Fig. 1.33 Wind/photovoltaic system/diesel generators with battery storage under MATLAB/ Simulink

References

1. Khan MJ, Iqbal MT (2009) Analysis of a small wind-hydrogen stand-alone hybrid energy system. Appl Energy 86(11):2429–2442. http://www.elsevier.com/inca/publications/store/4/0/5/8/9/1/index.htt, https://doi.org/10.1016/j.apenergy.2008.10.024
2. Karakoulidis K, Mavridis K, Bandekas DV, Adoniadis P, Potolias C, Vordos N (2011) Techno-economic analysis of a stand-alone hybrid photovoltaic-diesel-battery-fuel cell power system. Renew Energy 36(8):2238–2244. https://doi.org/10.1016/j.renene.2010.12.003
3. Eroglu M, Dursun E, Sevencan S, Song J, Yazici S, Kilic O (2011) A mobile renewable house using PV/wind/fuel cell hybrid power system. Int J Hydrogen Energy 36(13):7985–7992. https://doi.org/10.1016/j.ijhydene.2011.01.046
4. Ghasemi A, Asrari A, Zarif M, Abdelwahed S (2013) Techno-economic analysis of stand-alone hybrid photovoltaic-diesel-battery systems for rural electrification in eastern part of Iran—a step toward sustainable rural development. Renew Sustain Energy Rev 28:456–462. https://doi.org/10.1016/j.rser.2013.08.011
5. Genç G, Çelik M, Serdar Genç M (2012) Cost analysis of wind-electrolyzer-fuel cell system for energy demand in Pınarbaş-Kayseri. Int J Hydrogen Energy 37(17):12158–12166. https://doi.org/10.1016/j.ijhydene.2012.05.058
6. Shakya BD, Aye L, Musgrave P (2005) Technical feasibility and financial analysis of hybrid wind-photovoltaic system with hydrogen storage for Cooma. Int J Hydrogen Energy 30(1):9–20. https://doi.org/10.1016/j.ijhydene.2004.03.013
7. Bajpai P, Dash V (2012) Hybrid renewable energy systems for power generation in stand-alone applications: a review. Renew Sustain Energy Rev 16(5):2926–2939. https://doi.org/10.1016/j.rser.2012.02.009
8. Bernal-Agustín JL, Dufo-López R (2009) Simulation and optimization of stand-alone hybrid renewable energy systems. Renew Sustain Energy Rev 13(8):2111–2118. https://doi.org/10.1016/j.rser.2009.01.010
9. Deshmukh MK, Deshmukh SS (2008) Modeling of hybrid renewable energy systems. Renew Sustain Energy Rev 12(1):235–249. https://doi.org/10.1016/j.rser.2006.07.011
10. Bukar AL, Tan CW (2019) A review on stand-alone photovoltaic-wind energy system with fuel cell: system optimization and energy management strategy. J Clean Prod 221:73–88
11. Maheri A (2016) Effect of dispatch strategy on the performance of hybrid wind-PV battery-diesel-fuel cell systems. J Thermal Eng 2(4):820–825
12. Alramlawi M, Gabash A, Mohagheghi E, Li P (2018) Optimal operation of PV-battery-diesel microgrid for industrial loads under grid blackouts. In: Proceedings—2018 IEEE international conference on environment and electrical engineering and 2018 ieee industrial and commercial power systems Europe. EEEIC/I and CPS Europe 2018, art. no. 8493959
13. An LN, Tuan TQ (2018) Dynamic programming for optimal energy management of hybrid wind-PV-diesel-battery. Energies 11(11), art. no. 3045
14. Rodriguez-Gallegos CD, Gandhi O, Bieri M, Reindl T, Panda SK (2018) A diesel replacement strategy for off-grid systems based on progressive introduction of PV and batteries: an Indonesian case study. Appl Energy 229:1218–1232
15. Lu B, Shahidehpour M (2005) Short-term scheduling of battery in a grid-connected PV/battery system. IEEE Trans Power Syst 20(2):1053–1061. https://doi.org/10.1109/tpwrs.2005.846060
16. Wang C, Nehrir MH (2008) Power management of a stand-alone wind/photovoltaic/fuel cell energy system. IEEE Trans Energy Convers 23(3):957–967. https://doi.org/10.1109/tec.2007.914200
17. Chellali FB, Recioui A, Yaiche MR, Bentarzi H (2014) A hybrid wind/solar/diesel stand-alone system optimization for remote areas in Algeria. Int J Renew Energy Technol 5:12–24

18. Nelson DB, Nehrir MH, Wang C (2006) Unit sizing and cost analysis of stand-alone hybrid wind/PV/fuel cell power generation systems. Renew Energy 31(10):1641–1656. https://doi. org/10.1016/j.renene.2005.08.031

19. Ally CZ, Sun Y, De Jong ECW (2018) Impact of virtual inertia on increasing the hosting capacity of island diesel-PV ac-grid. In: Proceedings—2018 53rd International universities power engineering conference, UPEC 2018, art. no. 8541859

20. Rezkallah M, Singh S, Chandra, A (2017) Real-time hardware testing, control and performance analysis of hybrid cost-effective wind-PV-diesel standalone power generation system. In: 2017 IEEE industry applications society annual meeting, IAS 2017

21. Rezkallah M, Chandra A, Saad M, Tremblay M, Singh B, Singh S, Ibrahim H (2018) Composite control strategy for a PV-wind-diesel based off-grid power generation system supplying unbalanced non-linear loads. In: 2018 IEEE industry applications society annual meeting, IAS 2018 8544618

22. Jamalaiah A, Raju CP, Srinivasarao R (2017) Optimization and operation of a renewable energy based pv-fc-micro grid using homer. In: Proceedings of the international conference on inventive communication and computational technologies, ICICCT 2017, art. no. 7975238, pp 450–455. https://doi.org/10.1109/icicct.2017.7975238

23. Ntziachristos L, Kouridis C, Samaras Z, Pattas K (2005) A wind-power fuel-cell hybrid system study on the non-interconnected Aegean islands grid. Renew Energy 30(10):1471–1487. https://doi.org/10.1016/j.renene.2004.11.007

24. Khan A, Khan R (2018) Cost optimization of hybrid microgrid using solar PV, fuel cell and diesel generator in HOMER. In: ICECE 2018—2018 2nd International conference on energy conservation and efficiency, proceedings, art. no. 8554974, pp 14–18

25. Aziz AS, bin Tajuddin MFN, bin Adzman MR, Ramli MAM (2018) Feasibility analysis of PV/diesel/battery hybrid energy system using multi-year module. Int J Renew Energy Res 8 (4):1980–1993

26. Jamshidi M, Askarzadeh A (2019) Techno-economic analysis and size optimization of an off-grid hybrid photovoltaic, fuel cell and diesel generator system. Sustain Cities Soc 44:310–320

27. Djelailia O, Kelaiaia MS, Labar H, Necaibia S, Merad F (2019) Energy hybridization photovoltaic/diesel generator/pump storage hydroelectric management based on online optimal fuel consumption per kWh. Sustain Cities Soc 44:1–15

28. Tofani AF, Garniwa I, Fajry FR (2019) Techno-economic analysis of sea floating PV/diesel hybrid power plant with battery arrangement scheme for residential load at remote area in Indonesia (case study: Small Kei Island, South East Moluccas). In: Proceedings of 2018 international conference on electrical engineering and computer science, ICECOS 2018, art. no. 8605258, pp 243–246

29. Hosseinian H, Damghani H (2019) Ideal planning of a hybrid wind-PV-diesel microgrid framework with considerations for battery energy storage and uncertainty of renewable energy resources. In: 2019 IEEE 5th conference on knowledge based engineering and innovation, KBEI 2019, art. no. 8734947, pp 911–916

30. Qandil MD, Abbas AI, Qandil HD, Al-Haddad MR, Amano RS, A stand-alone hybrid photovoltaic, fuel cell, and battery system: case studies in Jordan

31. Gharibi M, Askarzadeh A (2019) Technical and economical bi-objective design of a grid-connected photovoltaic/diesel generator/fuel cell energy system. Sustainable Cities Soc 50:101575

32. Gharibi M, Askarzadeh A (2019) Size optimization of an off-grid hybrid system composed of photovoltaic and diesel generator subject to load variation factor. J Energy Storage 25:100814

33. Ghenai C, Bettayeb M, Brdjanin B, Hamid AK (2019) Hybrid solar PV/PEM fuel cell/diesel generator power system for cruise ship: a case study in Stockholm, Sweden. Case Stud Thermal Eng 14:100497

34. Adesanya AA, Schelly C (2019) Solar PV-diesel hybrid systems for the Nigerian private sector: an impact assessment. Energy Policy 196–207

35. García AM, Gallagher J, McNabola A, Poyato EC, Barrios PM, Rodríguez Díaz JR (2019) Comparing the environmental and economic impacts of on- or off-grid solar photovoltaics with traditional energy sources for rural irrigation systems. Renew Energy 895–904 (2019)
36. Yang D, Jiang C, Cai G, Huang N (2019) Optimal sizing of a wind/solar/battery/diesel hybrid microgrid based on typical scenarios considering meteorological variability. IET Renew Power Gener 13(9):1446–1455
37. Maritz J (2019) Optimized energy management strategies for campus hybrid PV-diesel systems during utility load shedding events. Processes 7(7):430
38. Movahediyan Z, Askarzadeh A (2019) A multiobjective approach for design of an off-grid PV/Diesel system considering reliability and cost. Environ Progr Sustain Energy 38 (4):13101
39. Ghenai C, Al-Ani I, Khalifeh F, Alamaari T, Hamid AK (2019) Design of solar PV/fuel cell/ diesel generator energy system for Dubai Ferry. In: Advances in science and engineering technology international conferences, ASET 2019. 8714292
40. Indriani SN, Anugrah P (2019) Design and economic analysis of off-grid PV system for diesel abatement in remote areas: case study of Indonesia. In: Proceedings—2018 international conference on control, electronics, renewable energy and communications, ICCEREC 2018, pp 129–135. 8712089
41. Singh B, Verma A, Chandra A, Al-Haddad K (2019) Implementation of solar PV-battery and diesel generator based electric vehicle charging station. In: Proceedings of 2018 IEEE international conference on power electronics, drives and energy systems, PEDES 2018. 8707673
42. Rekioua D (2014) Hybrid wind systems (book chapter), Issue 9781447164241. Green Energy and Technology, pp 163–183
43. Lu X, Qu Y, Wang Y, Qin C, Liu G (2018) A comprehensive review on hybrid power system for PEMFC-HEV: issues and strategies. Energy Convers Manag 171:1273–1291
44. Tamalouzt S, Benyahia N, Rekioua T, Rekioua D, Abdessemed R (2016) Wind turbine-DFIG/photovoltaic/fuel cell hybrid power sources system associated with hydrogen storage energy for micro-grid applications. In: Proceedings of 2015 IEEE international renewable and sustainable energy conference, IRSEC 2015, art. no. 7455060
45. Assaf J, Shabani B (2019) A novel hybrid renewable solar energy solution for continuous heat and power supply to standalone-alone applications with ultimate reliability and cost effectiveness. Renewable Energy 509–520
46. Assaf J, Shabani B (2018) Experimental study of a novel hybrid solar-thermal/PV-hydrogen system: towards 100% renewable heat and power supply to standalone applications. Energy 157:862–876
47. Farham H, Mohammadian L, Alipour H, Pouladi J (2018) Robust performance of photovoltaic/wind/grid based large electricity consumer. Sol Energy 174:923–932
48. Fathabadi H (2017) Novel standalone hybrid solar/wind/fuel cell power generation system for remote areas. Sol Energy 146:30–43
49. Handayani TP, Budiarto R, Hulukati SA, Gusa RF (2018) Development of a hybrid power supply control prototype for solar-powered water tank pumping system. In: Proceedings 2018 2nd international conference on green energy and applications, ICGEA 2018, pp 88–92
50. Ajiboye AA, Popoola SI, Atayero AA (2018) Hybrid renewable energy systems: opportunities and challenges in sub-Saharan Africa. In: Proceedings of the international conference on industrial engineering and operations management 2018 (SEP), pp 1110–1116
51. Gharibi M, Askarzadeh A (2019) Technical and economical bi-objective design of a grid-connected photovoltaic/diesel generator/fuel cell energy system. Sustain Cities Soc 50, art. no. 101575
52. Ghenai C, Bettayeb M (2019) Modelling and performance analysis of a stand-alone hybrid solar PV/fuel cell/diesel generator power system for university building. Energy 180–189
53. Ceran B (2019) The concept of use of PV/WT/FC hybrid power generation system for smoothing the energy profile of the consumer. Energy 853–865

54. Bensmail S, Rekioua D, Azzi H (2015) Study of hybrid photovoltaic/fuel cell system for stand-alone applications. Int J Hydrogen Energy 40(39):13820–13826
55. Mebarki NE, Rekioua T, Rekioua D (2016) Study of photovoltaic/hydrogen/ battery bank system to supply energy to an electric vehicle. In: Proceedings of 2015 IEEE international renewable and sustainable energy conference, IRSEC 2015. 7455055
56. Bartolucci L, Cordiner S, Mulone V, Santarelli M (2019) Short-therm forecasting method to improve the performance of a model predictive control strategy for a residential hybrid renewable energy system. Energy 997–1004
57. Bartolucci L, Cordiner S, Mulone V, Rossi JL (2019) Hybrid renewable energy systems for household ancillary services. Int J Electr Power Energy Syst 107:282–297
58. Tamalouzt S, Benyahia N, Rekioua T, Rekioua D, Abdessemed R (2016) Performances analysis of WT-DFIG with PV and fuel cell hybrid power sources system associated with hydrogen storage hybrid energy system. Int J Hydrogen Energy 41(45):21006–21021
59. Aissou S, Rekioua D, Mezzai N, Rekioua T, Bacha S (2015) Modeling and control of hybrid photovoltaic wind power system with battery storage. Energy Convers Manag 89:615–625
60. Bizon N, Thounthong P (2018) Fuel economy using the global optimization of the fuel cell hybrid power systems. Energy Convers Manag 173:665–678
61. Rekioua D, Bensmail S, Bettar N (2014) Development of hybrid photovoltaic-fuel cell system for stand-alone application. Int J Hydrogen Energy 39(3):1604–1611
62. Hassani H, Rekioua D, Aissou S, Bacha S (2019) Hybrid stand-alone photovoltaic/batteries/ fuel cells system for green cities. In: Proceedings of 2018 6th international renewable and sustainable energy conference, IRSEC 2018. 8702278
63. Zhang W, Maleki A, Rosen MA, Liu J (2019) Sizing a stand-alone solar-wind-hydrogen energy system using weather forecasting and a hybrid search optimization algorithm. Energy Convers Manag 609–621
64. Mengi OO, Altas IH (2015) A new energy management technique for PV/wind/grid renewable energy system. Int J Photoenergy 1–19
65. Balamurugan T, Manoharan S (2013) Optimal power flow management control for grid connected photovoltaic/wind turbine/diesel generator (GCPWD) hybrid system with batteries. Int J Renew Energy Res 3(4):819–826
66. Mancarella P (2014) Multi-energy systems: An overview of concepts and evaluation models. Energy 1–35
67. Papadopoulos V, Desmet J, Knockaert J, Develder C (2018) Improving the utilization factor of a PEM electrolyzer powered by a 15 MW PV park by combining wind power and battery storage—feasibility study. Int J Hydrogen Energy 43(34):16468–16478
68. Devrim Y, Pehlivanoglu K (2015) Design of a hybrid photovoltaic-electrolyzer-PEM fuel cell system for developing solar model. Physica Status Solidi (C) Curr Top Solid State Phys 12(9–11):1256–1261
69. Devrim Y, Bilir L (2016) Performance investigation of a wind turbine–solar photovoltaic panels–fuel cell hybrid system installed at İncek region—Ankara, Turkey. Energy Convers Manag 126:759–766
70. Dharavath R, Raglend IJ (2019) Intelligent controller based solar photovoltaic with battery storage, fuel cell integration for power conditioning. Int J Renew Energy Res 9(2):859–867
71. Wies RW, Johnson RA, Agrawal AN, Chubb JJ (2004) Economic analysis and environmental impacts of a PV with diesel-battery system for remote villages. In: 2004 IEEE power engineering society general meeting, vol. 2, pp 1898–1904. ISBN 0780384652; 978-078038465-1
72. Malla SG, Bhende CN (2014) Enhanced operation of stand-alone "photovoltaic-Diesel Generator-Battery" system. Electr Power Syst Res 107:250–257. https://doi.org/10.1016/j. epsr.2013.10.009
73. Krishan O, Sathans S (2016) Frequency regulation in a standalone wind-diesel hybrid power system using pitch-angle controller. In: Proceedings of the 10th INDIACom; 2016 3rd international conference on computing for sustainable global development, INDIACom 2016, art. no. 7724444, pp 1148–1152. ISBN 978-938054419-9

74. Olatomiwa L, Mekhilef S, Huda ASN, Sanusi K (2015) Techno-economic analysis of hybrid PV–diesel–battery and PV–wind–diesel–battery power systems for mobile BTS: the way forward for rural development. Energy Sci Eng 3(4):271–285. http://onlinelibrary.wiley.com/journal/10.1002/(ISSN)2050-0505, https://doi.org/10.1002/ese3.71

75. Qandil MD, Abbas AI, Qandil HD, Al-Haddad MR, Amano RS (2019) A stand-alone hybrid photovoltaic, fuel cell, and battery system: case studies in Jordan. J Energy Resour Technol Trans ASME 141(11):111201

76. Mishra PP, Fathy HK (2019) Achieving self-balancing by design in photovoltaic energy storage systems. IEEE Trans Control Syst Technol 27(3), art. no. 8327870, 1151–1164

77. Kosmadakis IE, Elmasides C, Eleftheriou D, Tsagarakis KP (2019) A techno-economic analysis of a pv-battery system in Greece. Energies 12(7), art. no. 1357

78. Kumar S, Singh B (2019) Seamless operation and control of single-phase hybrid PV-BES-utility synchronized system. IEEE Trans Ind Appl 55(2), art. no. 8496838, 1072–1082

79. Jing W, Lai CH, Ling DKX, Wong WSH, Wong MLD (2019) Battery lifetime enhancement via smart hybrid energy storage plug-in module in standalone photovoltaic power system. J Energy Storage 21:586–598

80. Liu J, Chen X, Cao S, Yang H (2019) Overview on hybrid solar photovoltaic-electrical energy storage technologies for power supply to buildings. Energy Convers Manag 103–121

81. Nengroo SH, Ali MU, Zafar A, Hussain S, Murtaza T, Alvi MJ, Raghavendra KVG, Kim HJ (2019) An optimized methodology for a hybrid photo-voltaic and energy storage system connected to a low-voltage grid. Electronics (Switzerland) 8(2), art. no. 176

82. Tayab UB, Yang F, El-Hendawi M, Lu J (2019) Energy management system for a grid-connected microgrid with photovoltaic and battery energy storage system. In: ANZCC 2018—2018 Australian and New Zealand control conference, art. no. 8606557, pp 141–144

83. Uctug FG, Azapagic A (2018) Environmental impacts of small-scale hybrid energy systems: coupling solar photovoltaics and lithium-ion batteries. Sci Total Environ 643:1579–1589

84. Spataru S, Martins J, Stroe D.-I, Sera D (2018) Test platform for photovoltaic systems with integrated battery energy storage applications. In: 2018 IEEE 7th world conference on photovoltaic energy conversion, WCPEC 2018—a joint conference of 45th IEEE PVSC, 28th PVSEC and 34th EU PVSEC, art. no. 8548122, pp 638–643

85. Martins J, Spataru S, Sera D, Stroe D-I, Lashab A (2019) Comparative study of ramp-rate control algorithms for PV with energy storage systems. Energies 12(7), art. no. 1342

86. Yi Z, Dong W, Etemadi AH (2018) A unified control and power management scheme for PV-battery-based hybrid microgrids for both grid-connected and islanded modes. IEEE Trans Smart Grid 9(6), art. no. 7917272, 5975–5985

87. Mahmood H, Jiang J (2018) Autonomous coordination of multiple PV/Battery hybrid units in islanded microgrids. In: IEEE Trans Smart Grid 9(6), art. no. 7935449, 6359–6368

88. Ye Y, Garg P, Sharma R (2013) Development and demonstration of power management of hybrid energy storage for PV integration. In: 2013 4th IEEE/PES innovative smart grid technologies Europe, ISGT Europe 2013, art. no. 6695470

89. Esan AB, Agbetuyi AF, Oghorada O, Ogbeide K, Awelewa AA, Afolabi AE (2019) Reliability assessments of an islanded hybrid PV-diesel-battery system for a typical rural community in Nigeria. Heliyon 5(5), art. no. e01632

90. Vargas U, Lazaroiu GC, Tironi E, Ramirez A (2019) Harmonic modeling and simulation of a stand-alone photovoltaic-battery-supercapacitor hybrid system. Int J Electr Power Energy Syst 105:70–78

91. Sedaghati R, Shakarami MR (2019) A novel control strategy and power management of hybrid PV/FC/SC/battery renewable power system-based grid-connected microgrid. Sustain Cities Soc 44:830–843

92. Liao J, Jiang Y, Li J, Liao Y, Du H, Zhu W, Zhang L (2018) An improved energy management strategy of hybrid photovoltaic/battery/fuel cell system for stratospheric airship. Acta Astronaut 152:727–739

93. Zheng H, Li S, Zang C, Zheng W (2013) Coordinated control for grid integration of PV array, battery storage, and supercapacitor. In: IEEE power and energy society general meeting, art. no. 6672725

94. Movahediyan Z, Askarzadeh A (2019) A multiobjective approach for design of an off-grid PV/diesel system considering reliability and cost.: Environ Prog Sustain Energy 38(4), art. no. 13101

95. Murugesan C, Marimuthu CN (2019) Cost optimization of PV-diesel systems in nanogrid using L J Cuckoo search and its application in mobile towers. Mobile Netw Appl 24(2): 340–349

96. Arabzadeh Saheli M, Fazelpour F, Soltani N, Rosen MA (2019) Performance analysis of a photovoltaic/wind/diesel hybrid power generation system for domestic utilization in Winnipeg, Manitoba, Canada. Environ Prog Sustain Energy 38(2):548–562

97. Tu T, Rajarathnam GP, Vassallo AM (2019) Optimization of a stand-alone photovoltaic–wind–diesel–battery system with multi-layered demand scheduling. Renew Energy 131:333–347

98. Agarwal N, Kumar A, Varun (2013) Optimization of grid independent hybrid PV-diesel-battery system for power generation in remote villages of Uttar Pradesh, India. Energy Sustain Dev 17(3):210–219. http://www.elsevier.com, https://doi.org/10.1016/j.esd.2013.02.002

99. Agarwala N, Varun KA (2012) Sizing analysis and cost optimization of hybrid solar-diesel-battery based electric power generation system using simulated annealing technique. Distrib Gener Altern Energy J 27(3):26–51. https://doi.org/10.1080/21563306.2012.10531122

100. Ashari M, Nayar CV (1999) An optimum dispatch strategy using set points for a photovoltaic (PV)-diesel-battery hybrid power system. Solar Energy 66 (1):1–9. www.elsevier.com/inca/publications/store/3/2/9/index.htt, https://doi.org/10.1016/s0038-092x(99)00016-x

101. Dursun B, Alboyaci B (2010) The contribution of wind-hydro pumped storage systems in meeting Turkey's electric energy demand. Renew Sustain Energy Rev 14(7):1979–1988. http://www.journals.elsevier.com/renewable-and-sustainable-energy-reviews/, https://doi.org/10.1016/j.rser.2010.03.030

102. Hossain M, Mekhilef S, Olatomiwa L (2017) Performance evaluation of a stand-alone PV-wind-diesel-battery hybrid system feasible for a large resort center in South China Sea, Malaysia. Sustain Cities Soc 28:358–366. http://www.elsevier.com/wps/find/journaldescription.cws_home/724360/description#description, https://doi.org/10.1016/j.scs.2016.10.008

103. Jing W, Lai CH, Wong WSH, Wong MLD (2018) A comprehensive study of battery-supercapacitor hybrid energy storage system for standalone PV power system in rural electrification. Appl Energy 224:340–356. http://www.elsevier.com/inca/publications/store/4/0/5/8/9/1/index.htt, https://doi.org/10.1016/j.apenergy.2018.04.106

104. Jurasz, J, Mikulik, J, Krzywda M, Ciapała B, Janowski M (2018) Integrating a wind- and solar-powered hybrid to the power system by coupling it with a hydroelectric power station with pumping installation. Energy 144:549–563. www.elsevier.com/inca/publications/store/4/8/3/, https://doi.org/10.1016/j.energy.2017.12.011

105. Nfah EM, Ngundam JM (2009) Feasibility of pico-hydro and photovoltaic hybrid power systems for remote villages in Cameroon. Renew Energy 34(6):1445–1450. https://doi.org/10.1016/j.renene.2008.10.019

106. Bakos GC (2002) Feasibility study of a hybrid wind/hydro power-system for low-cost electricity production. Appl Energy 72(3–4):599–608. http://www.elsevier.com/inca/publications/store/4/0/5/8/9/1/index.htt, https://doi.org/10.1016/s0306-2619(02)00045-4

107. Ma T, Yang H, Lu L, Peng J (2014) Technical feasibility study on a standalone hybrid solar-wind system with pumped hydro storage for a remote island in Hong Kong. Renew Energy 69:7–15. http://www.elsevier.com/inca/publications/store/9/6/9/index.htt, https://doi.org/10.1016/j.renene.2014.03.02

108. Petrakopoulou F, Robinson A, Loizidou M (2016) Simulation and analysis of a stand-alone solar-wind and pumped-storage hydropower plant. Energy 96:676–683. www.elsevier.com/inca/publications/store/4/8/3/, https://doi.org/10.1016/j.energy.2015.12.049

109. Chen C-L, Chen H-C, Lee J-Y (2016) Application of a generic superstructure-based formulation to the design of wind-pumped-storage hybrid systems on remote islands. Energy Convers Manag 111:339–351. https://doi.org/10.1016/j.enconman.2015.12.057

110. Notton G, Mistrushi, D, Stoyanov L, Berberi P (2017) Operation of a photovoltaic-wind plant with a hydro pumping-storage for electricity peak-shaving in an island context. Solar Energy 157:20–34. www.elsevier.com/inca/publications/store/3/2/9/index.htt, https://doi.org/10.1016/j.solener.2017.08.016

111. Khan MJ, Yadav AK, Mathew L (2017) Techno economic feasibility analysis of different combinations of PV-Wind-Diesel-Battery hybrid system for telecommunication applications in different cities of Punjab, India. Renew Sustain Energy Rev 76:577–607. https://doi.org/10.1016/j.rser.2017.03.076

112. Al-Ghussain L, Ahmed H, Haneef F (2018) Optimization of hybrid PV-wind system: case study Al-Tafilah cement factory, Jordan. Sustain Energy Technol Assess 30:24–36. http://www.journals.elsevier.com/sustainable-energy-technologies-and-assessments, https://doi.org/10.1016/j.seta.2018.08.008

113. Anoune K, Laknizi A, Bouya M, Astito A, Ben Abdellah A (2018) Sizing a PV-wind based hybrid system using deterministic approach. Energy Convers Manag 169:137–148. https://doi.org/10.1016/j.enconman.2018.05.034

114. Jurasz J, Ciapała B (2018) Solar–hydro hybrid power station as a way to smooth power output and increase water retention. Solar Energy 173:675–690. www.elsevier.com/inca/publications/store/3/2/9/index.htt, https://doi.org/10.1016/j.solener.2018.07.087

115. Jurasz J (2017) Modeling and forecasting energy flow between national power grid and a solar–wind–pumped-hydroelectricity (PV–WT–PSH) energy source. Energy Convers Manag 136:382–394. https://doi.org/10.1016/j.enconman.2017.01.032

116. Bizon N, Oproescu M, Raceanu M (2015) Efficient energy control strategies for a standalone renewable/fuel cell hybrid power source. Energy Convers Manag 90:93–110. https://doi.org/10.1016/j.enconman.2014.11.002

117. Rouholamini M, Mohammadian M (2015) Energy management of a grid-tied residential-scale hybrid renewable generation system incorporating fuel cell and electrolyzer. Energy Build 102:406–416. https://doi.org/10.1016/j.enbuild.2015.05.046

118. Tesfahunegn SG, Ulleberg Ø, Vie PJS, Undeland TM (2011) Optimal shifting of Photovoltaic and load fluctuations from fuel cell and electrolyzer to lead acid battery in a photovoltaic/hydrogen standalone power system for improved performance and life time. J Power Sour 196(23):10401–10414. https://doi.org/10.1016/j.jpowsour.2011.06.037

119. El-Shatter TF, Eskander MN, El-Hagry MT (2006) Energy flow and management of a hybrid wind/PV/fuel cell generation system. Energy Convers Manag 47(9–10):1264–1280. https://doi.org/10.1016/j.enconman.2005.06.022

120. Torreglosa JP, García P, Fernández LM, Jurado F (2015) Energy dispatching based on predictive controller of an off-grid wind turbine/photovoltaic/hydrogen/battery hybrid system. Renew Energy 74:326–336. http://www.elsevier.com/inca/publications/store/9/6/9/index.htt, https://doi.org/10.1016/j.renene.2014.08.010

121. Sanchez VM, Chavez-Ramirez AU, Duron-Torres SM, Hernandez J, Arriaga LG, Ramirez JM (2014) Techno-economical optimization based on swarm intelligence algorithm for a stand-alone wind-photovoltaic-hydrogen power system at south-east region of Mexico. Int J Hydrogen Energy 39(29):16646–16655. http://www.journals.elsevier.com/international-journal-of-hydrogen-energy/, https://doi.org/10.1016/j.ijhydene.2014.06.034

122. Rouholamini M, Mohammadian M (2016) Heuristic-based power management of a grid-connected hybrid energy system combined with hydrogen storage. Renew Energy Part A 96:354–365. http://www.journals.elsevier.com/renewable-and-sustainable-energy-reviews/, https://doi.org/10.1016/j.renene.2016.04.085

123. Tudu B, Mandal KK, Chakraborty N (2019) Optimal design and development of PV-wind-battery based nano-grid system: a field-on-laboratory demonstration. Front in Energy 13(2):269–283

124. Mohamad Izdin Hlal A, Ramachandaramurthya VK, Sanjeevikumar Padmanaban B, Hamid Reza Kaboli C, Aref Pouryekta A, Tuan Ab Rashid Bin Tuan Abdullah D (2019) NSGA-II and MOPSO based optimization for sizing of hybrid PV/wind/battery energy storage system. Int J Power Electron Drive Syst 10(1):463–478

125. Mirzapour F, Lakzaei M, Varamini G, Teimourian M, Ghadimi N (2019) A new prediction model of battery and wind-solar output in hybrid power system. J Ambient Intell Humaniz Comput 10(1):77–87

126. Sperstad IB, Korpas M (2019) Energy storage scheduling in distribution systems considering wind and photovoltaic generation uncertainties. Energies 12(7), art. no. 1231

127. Sandhu KS, Mahesh A (2016) A new approach of sizing battery energy storage system for smoothing the power fluctuations of a PV/wind hybrid system. Int J Energy Res 40(9):1221–1234

128. Dalton GJ, Lockington DA, Baldock TE (2008) Feasibility analysis of stand-alone renewable energy supply options for a large hotel. Renew Energy 33(7):1475–1490

129. Karasavvas KC (2008) Modular simulation of a hybrid power system with diesel, photovoltaic inverter and wind turbine generation. J Eng Sci Technol Rev 1(1):38–40

130. Seema Singh B (2019) Improved damped quadrature SOGI control algorithm for solar PV-hydro battery based microgrid. In: India international conference on power electronics, IICPE 2018-December. 8709339

131. Kewat S, Singh B, Hussain I (2017) Power management in PV-battery-hydro based standalone microgrid. IET Renew Power Gener

132. Seema Singh B (2018) PV-hydro-battery based standalone microgrid for rural electrification. In: 2018 5th IEEE Uttar Pradesh section international conference on electrical, electronics and computer engineering, UPCON 2018. 8597005

133. Tian C, Tian L, Li D, Lu X, Chang X (2016) Control strategy for tracking the output power of photovoltaic power generation based on hybrid energy storage system. Diangong Jishu Xuebao/Trans China Electrotechn Soc 31(14):75–83

134. Wang H, Jiancheng Z (2016) Research on charging/discharging control strategy of battery-super capacitor hybrid energy storage system in photovoltaic system. In: 2016 IEEE 8th international power electronics and motion control conference, IPEMC-ECCE Asia 2016, art. no. 7512723, pp 2694–2698

135. Cheng Z, Li Y, Xie Y, Qiu L, Dong B, Fan X (2015) Control strategy for hybrid energy storage of photovoltaic generation microgrid system with super capacitor. Dianwang Jishu/ Power Syst Technol 39(10):2739–2745

136. Lukic SM, Wirasingha SG, Rodriguez F, Cao J, Emadi A (2006) Power management of an ultracapacitor/battery hybrid energy storage system in an HEV. In: 2006 IEEE vehicle power and propulsion conference, VPPC 2006, art. no. 4211267. ISBN 1424401585; 978–142440158-1. https://doi.org/10.1109/vppc.2006.364357

137. Thounthong P, Raël S, Davat B (2009) Energy management of fuel cell/battery/ supercapacitor hybrid power source for vehicle applications. J Power Sour 193(1):376–385. https://doi.org/10.1016/j.jpowsour.2008.12.120

138. Tonkoski R, Lopes LAC, Turcotte D (2009) Active power curtailment of PV inverters in diesel hybrid mini-grids. In: 2009 IEEE electrical power and energy conference, EPEC 2009, art. no. 5420964

139. Colle S, Luna Abreu S, Rüther R (2004) Economic evaluation and optimization of hybrid diesel/photovoltaic systems integrated to utility grids. Solar Energy 76(1–3):295–299. www.elsevier.com/inca/publications/store/3/2/9/index.htt, https://doi.org/10.1016/j.solener.2003.08.008

140. Elhadidy MA, Shaahid SM (1999) Optimal sizing of battery storage for hybrid (wind + diesel) power systems. Renew Energy 18(1):77–86. http://www.journals.elsevier.com/renewable-and-sustainable-energy-reviews/, https://doi.org/10.1016/s0960-1481(98)00796-4

141. McGowan JG, Manwell JF, Connors SR (1988) Wind/diesel energy systems: review of design options and recent developments. Sol Energy 41(6):561–575. https://doi.org/10.1016/0038-092x(88)90059-x

142. Nayar CV, Phillips SJ, James WL, Pryor TL, Remmer D (1993) Novel wind/diesel/battery hybrid energy system. Sol Energy 51(1):65–78. https://doi.org/10.1016/0038-092x(93)90043-n

143. Elhadidy MA, Shaahid SM (1998) Feasibility of hybrid (wind + solar) power systems for Dhahran Saudi Arabia. In: World renewable energy congress, vol 5, Florence-Italy, pp 20–25

144. Phillips S, Nayar L (1988) Control and interfacing of photovoltaic/wind and diesel systems. National Energy Research Development and Demonstration Council End of Grant, Report No. NERDDP/EG/90/864. Department of Primary Industries and Energy, Canberra

145. Paska J, Biczel P (2005) Hybrid photovoltaic-fuel cell power plant. In: 2005 IEEE Russia Power Tech. PowerTech, art. no. 4524654

146. Nelson DB, Nehrir MH, Wang C (2005) Unit sizing of stand-alone hybrid wind/PV/fuel cell power generation systems. In: 2005 IEEE power engineering society general meeting, vol 3, pp 2116–2122

147. Karlis, A, Dokopoulos P (1996) Small power systems fed by hydro, photovoltaic, wind turbines and diesel generators. In: Proceedings of the IEEE international conference on electronics, circuits, and systems, vol 2, pp 1013–1016

148. Bhandari B, Poudel SR, Lee K-T, Ahn S-H (2014) Mathematical modeling of hybrid renewable energy system: a review on small hydro-solar-wind power generation. Int J Precis Eng Manufact Green Technol 1(2):157–173. http://www.springer.com/engineering/production+engineering/journal/40684, https://doi.org/10.1007/s40684-014-0021-4

149. Bhandari B, Poudel SR, Lee K-T, Ahn S-H (2014) Mathematical modeling of hybrid renewable energy system: A review on small hydro-solar-wind power generation. Int J Precis Eng Manufact Green Technol 1(2):157–173. http://www.springer.com/engineering/production+engineering/journal/40684, https://doi.org/10.1007/s40684-014-0021-4

150. Alvarez, SR, Ruiz AM, Oviedo JE (2017) Optimal design of a diesel-PV-wind system with batteries and hydro pumped storage in a Colombian community. In: 6th International conference on renewable energy research and applications, ICRERA 2017, 2017-January, pp 234–239

151. Kumar R, Gupta RA, Bansal AK (2013) Economic analysis and power management of a stand-alone wind/photovoltaic hybrid energy system using biogeography based optimization algorithm. Swarm Evolution Comput 8:33–43. https://doi.org/10.1016/j.swevo.2012.08.002

152. Li X, Hui D, Lai X, Yan T (2011) Power quality control in wind/fuel cell/battery/hydrogen electrolyzer hybrid micro-grid power system

153. Sanchez JA, Moreno N, Vazquez S, Carrasco JM, Galvan E, Batista C, Hurtado S, Costales G (2003) A 800 kW wind-diesel test bench based on the MADE AE-52 variable speed wind turbine. In: IECON proceedings (industrial electronics conference), vol 2, pp 1314–1319

154. Singh M, Chandra A (2009) Control of PMSG based variable speed wind-battery hybrid system in an isolated network. In: 2009 IEEE power and energy society general meeting, PES '09, art. no. 5275419

155. Cardenas R, Pena R, Clare J, Asher G (2003) Power smoothing in a variable speed wind-diesel system. In: PESC record—IEEE annual power electronics specialists conference, vol 2, pp 754–759

156. Elmitwally A, Rashed M (2011) Flexible operation strategy for an isolated PV-diesel microgrid without energy storage. IEEE Trans Energy Convers 26(1), art. no. 5648756, 235–244

157. Abedini A, Nikkhajoei H (2011) Dynamic model and control of a wind-turbine generator with energy storage. IET Renew Power Gener 5(1):67–78

158. Ko HS, Niimura T, Lee KY (2003) An intelligent controller for a remote wind-diesel power system—design and dynamic performance analysis. In: 2003 IEEE power engineering society general meeting, conference proceedings, vol 4, pp 2147–2152

159. Luu T, Nasiri A (2010) Power smoothing of doubly fed induction generator for wind turbine using ultracapacitors. In: IECON proceedings (industrial electronics conference), art. no. 5675040, pp 3293–3298

160. Gayathri NS, Senroy N, Kar IN (2017) Smoothing of wind power using flywheel energy storage system. IET Renew Power Gener 11(3):289–298

161. Mir AS, Senroy N (2018) Intelligently Controlled flywheel storage for wind power smoothing. In: IEEE power and energy society general meeting, 2018-August, art. no. 8585853

162. Li K, Xu H, Ma Q, Zhao J (2014) Hierarchy control of power quality for wind - Battery energy storage system. IET Power Electron 7(8):2123–2132. http://scitation.aip.org/dbt/dbt.jsp?KEY=IPEEBO, https://doi.org/10.1049/iet-pel.2013.0654

163. Iglesias IJ, García-Tabarés L, Agudo A, Cruz I, Arribas L (2000) Design and simulation of a stand-alone wind-diesel generator with a flywheel energy storage system to supply the required active and reactive power. In: PESC record—IEEE annual power electronics specialists conference, vol 3, pp 1381–1386. https://doi.org/10.1109/pesc.2000.880510

164. Hunter R, Elliot G (1994) Wind-diesel systems: a guide to the technology and its implementation. ISBN:0521434408; 978-052143440-9

165. Grudkowski TW, Dennis AJ, Meyer TG, Wawrzonek PH (1996) Flywheels for energy storage. SAMPE J 32(1):65–69

166. Suvire GO, Mercado PE (2012) Active power control of a flywheel energy storage system for wind energy applications. IET Renew Power Gener 6(1):9–16. https://doi.org/10.1049/iet-rpg.2010.0155

167. Taj TA, Hasanien HM, Alolah AI, Muyeen SM (2015) Transient stability enhancement of a grid connected wind farm using an adaptive neuro fuzzy controlled-flywheel energy storage system. In: IET Renew Power Gener 9(7):792–800. http://www.theiet.org/doi:10.1049/iet-rpg.2014.0345

168. Mandic G, Nasiri A, Ghotbi E, Muljadi E (2013) Lithium–ion capacitor energy storage integrated with variable speed wind turbines for power smoothing. In: IEEE J. Emerging Sel Top Power Electron 1(4), art. no. 6619430, 287–295

169. Abedini A, Nasiri A (2008) Applications of super capacitors for PMSG wind turbine power smoothing. In: IECON proceedings (industrial electronics conference), art. no. 4758497, pp 3347–3351

170. Howlader AM, Izumi Y, Uehara A, Urasaki N, Senjyu T, Yona A, Saber AY (2012) A minimal order observer based frequency control strategy for an integrated wind-battery-diesel power system. Energy 46(1):168–178. www.elsevier.com/inca/publications/store/4/8/3/doi:10.1016/j.energy.2012.08.039

171. Genç MS, Çelik M, Karasu I (2012) A review on wind energy and wind-hydrogen production in Turkey: a case study of hydrogen production via electrolysis system supplied by wind energy conversion system in Central Anatolian Turkey. Renew Sustain Energy Rev 16(9):6631–6646. https://doi.org/10.1016/j.rser.2012.08.011

172. Muyeen SM, Takahashi R, Murata T, Tamura J (2009) Integration of an energy capacitor system with a variable-speed wind generator. In: IEEE Trans Energy Convers 24(3):740–749. https://doi.org/10.1109/tec.2009.2025323

173. Abbey C, Joos G (2007) Supercapacitor energy storage for wind energy applications. IEEE Trans Ind Appl 43(3):769–776. https://doi.org/10.1109/tia.2007.895768

174. Teleke S, Baran ME, Bhattacharya S, Huang AQ (2010) Optimal control of battery energy storage for wind farm dispatching. IEEE Trans Energy Convers 25(3), art. no. 5432995, 787–794. https://doi.org/10.1109/tec.2010.2041550

175. Leclercq L, Robyns B, Grave J-M (2003) Control based on fuzzy logic of a flywheel energy storage system associated with wind and diesel generators. Math Comput Simul 63(3–5):271–280. https://doi.org/10.1016/s0378-4754(03)00075-

176. Chedid RB, Karaki SH, El-Chamali C (2000) Adaptive fuzzy control for wind-diesel weak power systems. IEEE Trans Energy Convers 15(1):71–78. https://doi.org/10.1109/60. 849119

177. Cárdenas R, Peña R, Asher G, Clare J (2001) Control strategies for enhanced power smoothing in wind energy systems using a flywheel driven by a vector-controlled induction machine. IEEE Trans Ind Electron 48(3):625–635. https://doi.org/10.1109/41.925590

178. Iqbal MT (2003) Modeling and control of a wind fuel cell hybrid energy system. Renew Energy 28(2):223–237. https://doi.org/10.1016/s0960-1481(02)00016-2

179. Ogueke NV, Ikpamezie II, Anyanwu EE (2016) The potential of a small hydro/photovoltaic hybrid system for electricity generation in FUTO, Nigeria. Int J Ambient Energy 37(3):256–265. http://www.tandfonline.com/toc/taen20/current, https://doi.org/10.1080/01430750. 2014.952841

180. An Y, Fang W, Ming B, Huang Q (2015) Theories and methodology of complementary hydro/photovoltaic operation: applications to short-term scheduling. J Renew Sustain Energy 7(6), art. no. 063133. http://scitation.aip.org/content/aip/journal/jrse, https://doi.org/10.1063/1.4939056

181. Marinescu C, Serban I (2011) Robust frequency control for a wind/hydro autonomous microgrid. In: 2011 IEEE PES Trondheim PowerTech: the power of technology for a sustainable society, POWERTECH 2011, art. no. 6019248

182. Keskin Arabul F, Arabul AY, Kumru CF, Boynuegri AR (2017) Providing energy management of a fuel cell–battery–wind turbine–solar panel hybrid off grid smart home system. Int J Hydrogen Energy 42(43):26906–26913

183. Bahtiyar D, Ercan A (2019) An investigation on wind/PV/fuel cell/battery hybrid renewable energy system for nursing home in Istanbul. Proc Inst Mech Eng Part A: J Power Energy 0 (0):1–10

184. Nojavan S, Majidi M, Zare K (2017) Risk-based optimal performance of a PV/fuel cell/ battery/grid hybrid energy system using information gap decision theory in the presence of demand response program. Int J Hydrogen Energy 42:11857–11867

185. Silva SB, de Oliveira MAG, Severino MM (2010) Economic evaluation and optimization of a photovoltaic–fuel cell–batteries hybrid system for use in the Brazilian Amazon. Energy Policy 38:6713–6723

186. Maleki A (2018) Modeling and optimum design of an off-grid PV/WT/FC/diesel hybrid system considering different fuel prices. Int J Low-Carbon Technol 13(2):140–147

187. Abdullah MO, Yung VC, Anyi M, Othman AK, Ab. Hamid KB, Tarawe J (2010) Review and comparison study of hybrid diesel/solar/hydro/fuel cell energy schemes for a rural ICT Telecenter. Energy 35(2):639–646. www.elsevier.com/inca/publications/store/4/8/3/doi:10. 1016/j.energy.2009.10.035

188. Maleki A, Pourfayaz F (2015) Sizing of stand-alone photovoltaic/wind/diesel system with battery and fuel cell storage devices by harmony search algorithm. J Energy Storage 2:30–42. http://www.journals.elsevier.com/journal-of-energy-storage/, https://doi.org/10.1016/j. est.2015.05.006

189. Ramli MAM, Hiendro A, Twaha S (2015) Economic analysis of PV/diesel hybrid system with flywheel energy storage. Renew Energy 78:398–405. http://www.journals.elsevier.com/ renewable-and-sustainable-energy-reviews/, https://doi.org/10.1016/j.renene.2015.01.026

190. Raja Manickam P, Muniyaraj M, Kaliraj V, Shanmugapriya ST Model of hybrid solar wind diesel fuel cell power system. IOSR J. Electr. Electron. Eng. (IOSR-JEEE), pp 12–17

191. Abdullah MO, Yung VC, Anyi M, Othman AK, Ab. Hamid KB, Tarawe J (2010) Review and comparison study of hybrid diesel/solar/hydro/fuel cell energy schemes for a rural ICT Telecenter. Energy 35(2):639–646

192. Nigussie T, Bogale W, Bekele F, Dribssa E (2017) Feasibility study for power generation using off- grid energy system from micro hydro-PV-diesel generator-battery for rural area of Ethiopia: the case of Melkey Hera village, Western Ethiopia. AIMS Energy 5(4):667–690

193. Perrin M, Malbranche P, Mattera F, Simonin L, Sauer DU, Lailler P, Joessen A, Willer B, Dahlen M, Ruddell A, Bruesewitz M (2003) Evaluation and perspectives of storage technologies for PV electricity. In: Proceedings of the 3rd world conference on photovoltaic energy conversion, C, pp 2194–2197

194. Lee D-J, Wang L (2008) Small-signal stability analysis of an autonomous hybrid renewable energy power generation/energy storage system part I: time-domain simulations. IEEE Trans Energy Convers 23(1):311–320

195. Kaldellis JK (2010) Stand-alone and hybrid wind energy systems: technology, energy storage and applications. In: Stand-alone and hybrid wind energy systems: technology, energy storage and applications, pp 1–554

196. Luta DN, Raji AK (2019) Optimal sizing of hybrid fuel cell-supercapacitor storage system for off-grid renewable applications. Energy 530–540

197. Abdelkader A, Rabeh A, Mohamed Ali D, Mohamed J (2018) Multi-objective genetic algorithm based sizing optimization of a stand-alone wind/PV power supply system with enhanced battery/supercapacitor hybrid energy storage. Energy 163:351–363

198. Thounthong P, Sikkabut S, Mungporn P, Sethakul P, Pierfederici S, Davat B (2013) Differential flatness based-control of fuel cell/photovoltaic/wind turbine/supercapacitor hybrid power plant. In: 4th International conference on clean electrical power: renewable energy resources impact, ICCEP 2013, art. no. 6587005, pp 298–305

199. Makbul A, Ayong H, Ssennoga T (2015) Economic analysis of PV/diesel hybrid system with flywheel energy storage. Renewable Energy 78:398–405

200. Lal DK, Dash BB, Akella AK (2011) Optimization of PV/Wind/Micro-Hydro/diesel hybrid power system in homer for the study area. Int J Electr Eng Inform 3(3):307–325

201. Canales FA, Beluco A, Mendes CAB (2015) A comparative study of a wind hydro hybrid system with water storage capacity: conventional reservoir or pumped storage plant? J Energy Storage 4:96–105

202. Sri Revathi B, Mahalingam P, Gonzalez-Longatt F (2019) Interleaved high gain DC-DC converter for integrating solar PV source to DC bus. Sol Energy 188:924–934

203. Croci L, Martinez A, Coirault P, Champenois G (2012) Control strategy for photovoltaic-wind hybrid system using sliding mode control and linear parameter varying feedback. In: ICIT 2012 (Athena, Greece), 19–21 Mar 2012

204. Rekioua D, Zaouche F, Hassani H, Rekioua T, Bacha S (2019) Modeling and fuzzy logic control of a stand-alone photovoltaic system with battery storage. Turkish J Electromech Energy 4(1)

Chapter 2
Power Electronics in Hybrid Renewable Energies Systems

2.1 Introduction

Renewable energy generators are almost always associated with power electronics. The configuration of a hybrid system essentially depends on the choice of static converters. Indeed, for a DC bus architecture, all generators must be connected in series with the inverter in order to supply alternative loads. On the other hand, in the case of an AC bus configuration, each converter will be associated with its generator in such a way as to supply the load independently and simultaneously with the other converters. In the mixed DC/AC configuration, the converters located between two buses (the rectifier and inverter) can be replaced by a bidirectional converter In general case, the choice of a converter depends on the nature of the source and of the load. In this chapter, the different converters used in photovoltaic, wind power and storage systems are presented with some examples under MATLAB/Simulink, as well as some control techniques applied to static converters.

2.2 Power Electronic Converters

Power electronic converters are made up of power semiconductors switches (diodes, transistors …) and passive components. Passive components are mainly inductive (inductors, transformers, coupled inductors) or capacitive (capacitors). The choice of a converter depends on the nature of the source and of the load (Fig. 2.1) [1–21].

© Springer Nature Switzerland AG 2020

D. Rekioua, *Hybrid Renewable Energy Systems*, Green Energy and Technology,

https://doi.org/10.1007/978-3-030-34021-6_2

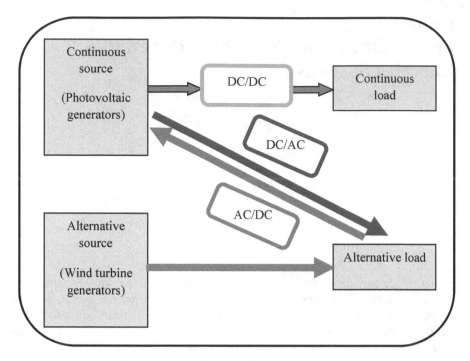

Fig. 2.1 Various static converters depending on loads

2.3   Power Converters in PV Systems

2.3.1   DC/DC Converters

In PV systems, generally, DC/DC converter operates as boost converter. It is placed between the source (PV generator) and the load (Fig. 2.2) [3, 4].

The correct control of the duty cycle allows maximum power in the load (Fig. 2.3).

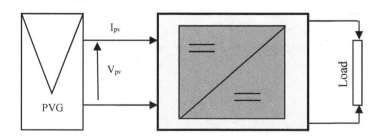

Fig. 2.2 DC/DC converter in PV system

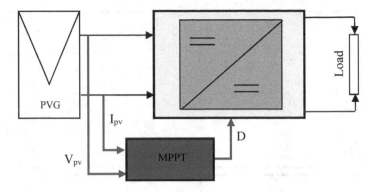

Fig. 2.3 DC/DC converter with MPPT controller in standalone PV system

Voltage converters can be divided into three types according to the position of the switch and chopper (Figs. 2.4, 2.5, 2.6):

- Buck converter:

When Kirchhoff's law is applied on the above circuit, the following equations are obtained [22–24]:

$$I_{c1}(t) = \frac{C_1 dV_{in}(t)}{dt} = I_{in}(t) - I_L(t) \tag{2.1}$$

$$I_{c2}(t) = \frac{C_2 dV_{out}(t)}{dt} = I_{in}(t) - I_{out}(t) \tag{2.2}$$

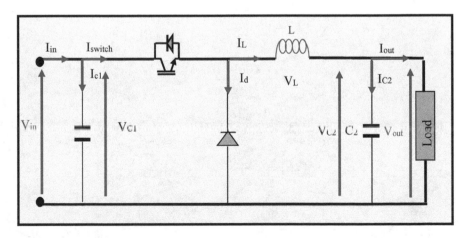

Fig. 2.4 Buck converter

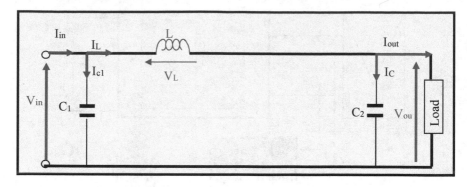

Fig. 2.5 Configuration between time 0 and $D.T$

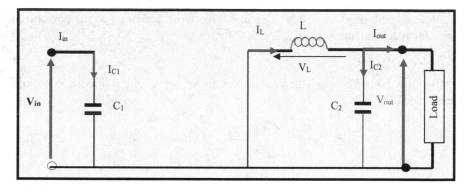

Fig. 2.6 Configuration between time $D.T$ and T

$$V_L(t) = \frac{L.dI_L(t)}{dt} = V_{in}(t) - V_{out}(t) \tag{2.3}$$

$$I_{c1}(t) = \frac{C_1 dV_{in}(t)}{dt} = I_{in}(t) \tag{2.4}$$

$$I_{c2}(t) = \frac{C_2 dV_{out}(t)}{dt} = I_L(t) - I_{out}(t) \tag{2.5}$$

$$V_L(t) = \frac{L dI_L(t)}{dt} = -V_{out}(t) \tag{2.6}$$

The average model is given as:

$$
i_{C1} = I_{PV} - I_L
$$
$$
i_{C2} = -I_{\text{load}} \tag{2.7}
$$
$$
v_L = V_{PV}
$$

- Boost converter

The DC/DC boost converter is used due to the higher efficiency of conversion and its simple structure (Fig. 2.7).

The first configuration is shown in Fig. 2.8.

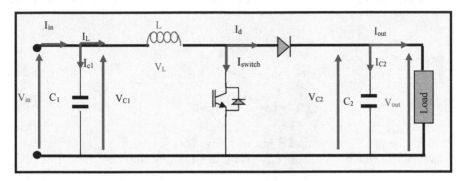

Fig. 2.7 Boost converter

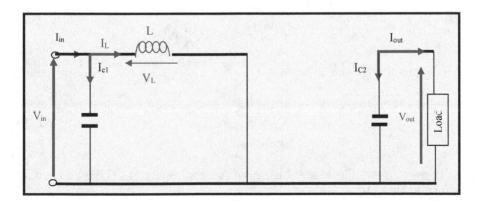

Fig. 2.8 First boost configuration

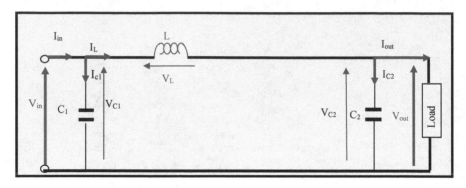

Fig. 2.9 Second boost configuration

For the interval $D.T$, Kirchhoff's law leads to the following set of expressions:

$$i_{C1}(t) = C_1 \frac{dv_{c1}(t)}{dt} = i_{PV}(t) - i_L(t)$$
$$i_{C2}(t) = C_2 \frac{dv_{c2}(t)}{dt} = -i_{\text{load}}(t) \qquad (2.8)$$
$$v_L(t) = L \frac{di_L(t)}{dt} = v_{PV}(t)$$

And, the first configuration is given in Fig. 2.9.
When the switch is on, the corresponding equations are:

$$i_{C1}(t) = C_1 \frac{dv_{c1}(t)}{dt} = i_{PV}(t) - i_L(t)$$
$$i_{C2}(t) = C_2 \frac{dv_{c2}(t)}{dt} = i_L(t) - i_{\text{load}}(t) \qquad (2.9)$$
$$v_L(t) = L \frac{di_L(t)}{dt} = v_{PV}(t) - v_{\text{load}}(t)$$

The average model is given as:

$$i_{C1} = I_{PV} - I_L$$
$$i_{C2} = I_L - I_{\text{load}} \qquad (2.10)$$
$$v_L = V_{PV} - V_{\text{load}}$$

Boost converter block model has been designed under MATLAB/Simulink as shown in Fig. 2.10.
The regulation output voltage can be made as (Fig. 2.11).

- **Design of the boost converter:**

The design of the boost converter involves calculating the value of inductance L and the capacity C [12].

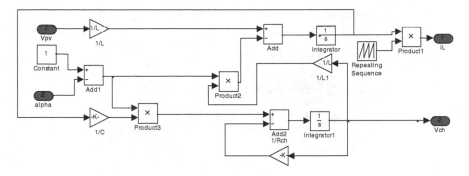

Fig. 2.10 Simulink model of the boost converter

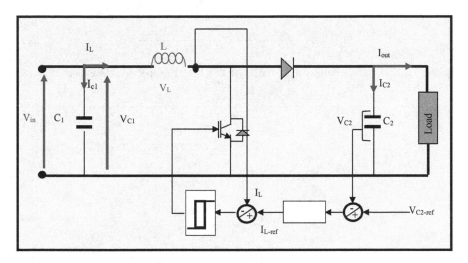

Fig. 2.11 Output voltage regulation in a boost converter

$$V_L = V_{PV} = L\frac{\mathrm{d}i_L(t)}{\mathrm{d}t} \tag{2.11}$$

$$i_L = \int_0^t \frac{V_{PV}}{L}\,\mathrm{d}t = \frac{V_{PV}}{L}t + I_{L\min} \tag{2.12}$$

At $t = D.T$

$$i_L = \frac{V_{PV}}{L}D.T + I_{L\min} = I_{L\max} \tag{2.13}$$

The current ripple is:

$$I_{Lmax} - I_{Lmin} = \Delta I_L = \frac{V_{PV}}{L} D.T = \frac{(1-D)V_{load}}{L} DT$$

The current ripple is maximal when:

$$\frac{d(\Delta I_L)}{dt} = 0 \Rightarrow \frac{(1-2.D)T.V_{PV}}{L} = 0 \Rightarrow D = \frac{1}{2}$$

Then:

$$(\Delta I_L)_{max} = \frac{1/4\ T.V_{PV}}{L} \tag{2.14}$$

$$L = \frac{1/4\ T.V_{PV}}{(\Delta I_L)_{max}} = \frac{V_{PV}.D}{2.f_{swi}.(\Delta I_L)_{max}} \tag{2.15}$$

where: f_{swi} is the switching frequency of the boost converter, D is the duty ratio (Fig. 2.12),

$$V_C = \frac{1}{C}\int i_c dt \tag{2.16}$$

$$i_c = -i_{Load} = -\frac{V_{Load}}{R_{Load}} \tag{2.17}$$

$$V_c = \frac{1}{C}\int -\frac{V_{Load}}{R_{Load}} dt \tag{2.18}$$

$$V_c = V_{Cmax} - \frac{V_c}{C.R_{Load}} t \tag{2.19}$$

At $t = D.T$:

$$V_c = V_{Cmax} - \frac{V_c}{CR_{Load}} D.T = V_{Cmin} \tag{2.20}$$

Fig. 2.12 Current ripple waveforms

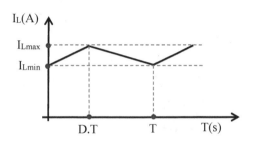

Fig. 2.13 Voltage ripple

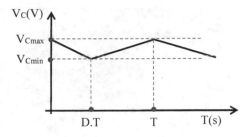

The voltage ripple is:

$$V_{\text{Cmax}} - V_{\text{Cmin}} = \Delta V_C = \frac{V_{Load}}{C.R_{Load}} D.T \tag{2.21}$$

$$\Delta V_C = \frac{1/2\ T.V_{\text{Load}}}{C.R_{\text{Load}}} \tag{2.22}$$

Thus (Fig. 2.13):

$$C = \frac{1/2\ T.V_{\text{Load}}}{\Delta V_C.R_{\text{Load}}} = \frac{D.V_{\text{Load}}}{f_{\text{swi}}.\Delta V_C.R_{\text{Load}}} \tag{2.23}$$

The number of coil turns can be written as:

$$N = \sqrt{L/AL} \tag{2.24}$$

With, L: the inductance (mH), N: number of coil turns, AL: Toroidal inductor (nH/tr^2).

A realized boost converter is shown in Fig. 2.14 [22].

To test the boost converter performances, we have made some tests on a resistive load (Table 2.1).

Figure 2.15 shows measured self-voltage and the output capacity voltage waveforms of the boost converter.

- Buck–Boost converter

The two inductors L_1 and L_2 can be combined to one inductor L and the capacitor C_1 can be removed. It is obtained the following scheme (Fig. 2.17) with the same operation as in the previous figure (Fig. 2.16). It is a non-inverting buck–boost converter [25].

This scheme can be reduced by limiting the number of switches, but it will reverse the polarity of the output voltage (Fig. 2.18).

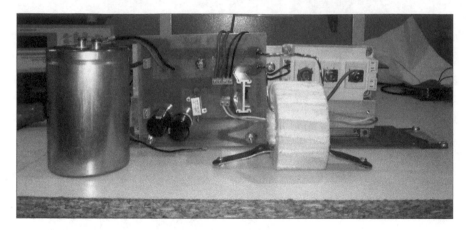

Fig. 2.14 Developed boost converter

Table 2.1 Boost converter experimental results under different tests

Tests	Input variables			Output variables			Efficiency η (%)
	I_{in} (A)	V_{in} (V)	P_{in} (W)	I_{out} (A)	V_{out} (V)	P_{out} (W)	
1	3.8	53	201.5	2	82.25	164.5	81.69
2	3.83	72.75	278.75	2	113.7	227.4	81.57
3	3.95	97.97	387	2	157.65	315.3	81.46

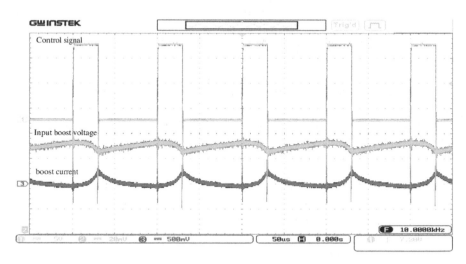

Fig. 2.15 Measured boost voltage, current, power and control signal

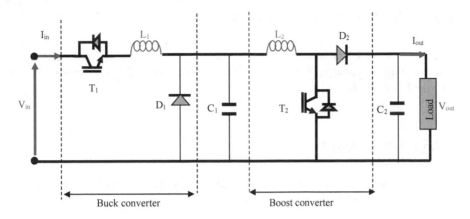

Fig. 2.16 Buck–boost converter

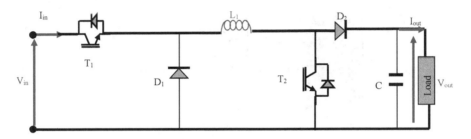

Fig. 2.17 Non-inverting buck–boost converter

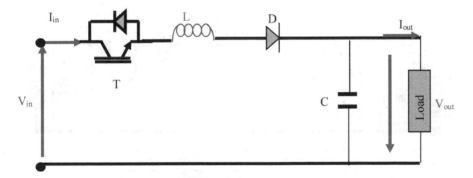

Fig. 2.18 Inverting buck–boost converter

2.3.2 Application in PV Systems

The output of a PV generator is connected directly to a boost converter with a constant output voltage (Fig. 2.19a). The circuit in Fig. 2.19b consists of two cascaded converters [18].

2.3.3 Application in Wind System

A DC/DC boost converter interfaces with wind system to maintain constant output voltage (Fig. 2.20) [6].

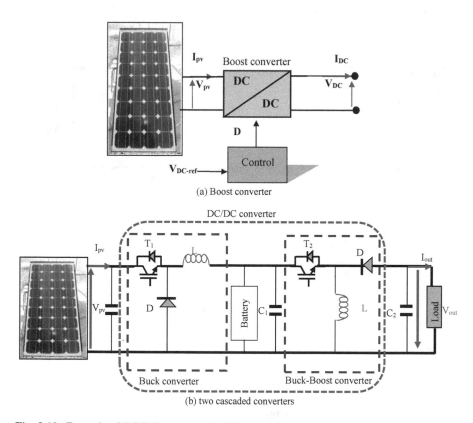

Fig. 2.19 Example of DC/DC converters in PV system

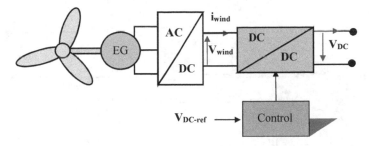

Fig. 2.20 Example of DC/DC converter in wind system

2.3.4 Application in Storage Systems

The bidirectional DC/DC converter is an essential element for connecting storage systems between source and load for continuous power exchange (Fig. 2.21) [5, 11–14, 17].

2.3.5 Application to Hybrid PV/Wind System with Battery Storage

Hybrid PV/Wind system includes a photovoltaic generator with a DC/DC converter, a wind turbine that ensures the conversion of wind energy into electricity. Both of the two subsystems are connected to a DC bus. The storage is provided by batteries. The supplied load is alternatively connected through an inverter (Fig. 2.22) [18, 19].

In the case of a photovoltaic system with hybrid storage (batteries, fuel cells), we have the following structure with the different DC/DC converters that are used (Fig. 2.23) [20].

The buck–boost model under MATLAB/Simulink is as (Fig. 2.24).

2.3.6 DC/AC Converters

An inverter converts a DC voltage to an AC sinusoidal voltage (Fig. 2.25).

Figure 2.26 shows the electrical schematic of three-phase inverter. The DC/AC converter comprises six Insulated Gate Bipolar Transistors (IGBTs) to control the three-phase load. The purpose of this converter is to manage the amplitude and frequency of the stator voltages. The stator phase voltages are expressed using [7, 8]:

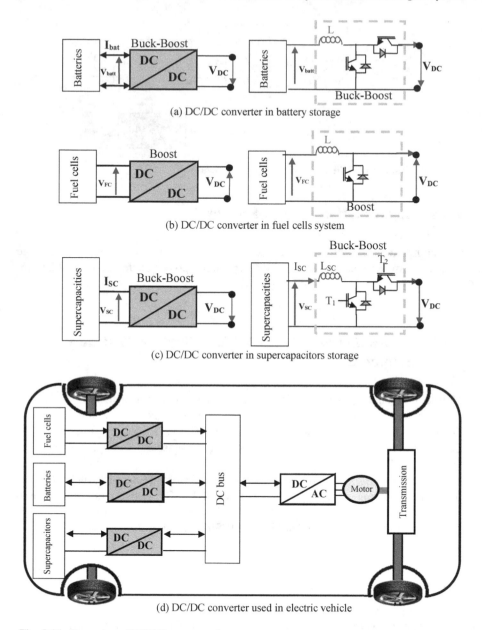

(a) DC/DC converter in battery storage

(b) DC/DC converter in fuel cells system

(c) DC/DC converter in supercapacitors storage

(d) DC/DC converter used in electric vehicle

Fig. 2.21 Examples of DC/DC converter in storage systems

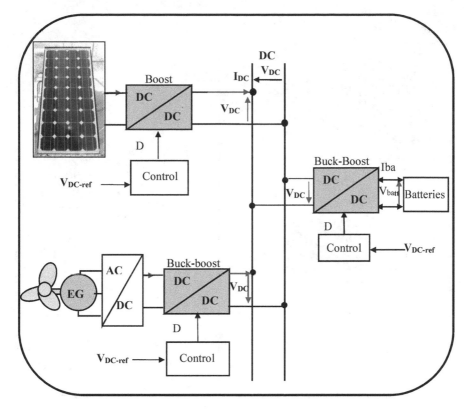

Fig. 2.22 DC/DC converters in hybrid PV/wind system with battery storage

$$\begin{bmatrix} v_{as} \\ v_{bs} \\ v_{cs} \end{bmatrix} = \frac{V_{DC}}{3} \begin{bmatrix} 2 & -1 & -1 \\ -1 & 2 & -1 \\ -1 & -1 & 2 \end{bmatrix} \cdot \begin{bmatrix} S_a \\ S_b \\ S_c \end{bmatrix} \qquad (2.25)$$

where: v_{as}, v_{bs} and v_{cs} are the three-phase stator voltages, V_{DC}, the DC link voltage, and $S_{(a,b,c)}$ are the switching function.

2.3.6.1 PWM Strategy

A natural PWM strategy can be used to drive the full-bridge DC/AC inverter (Fig. 2.26). The three-phase inverter consists of three legs, one for each phase. The control signals are generated by modulating three low-frequency sinusoidal (v_a, v_b, v_c) with a common high-frequency triangular carrier wave (V_p). For the inverter control, the switches logical state is used as a function of the control signals as [2]:

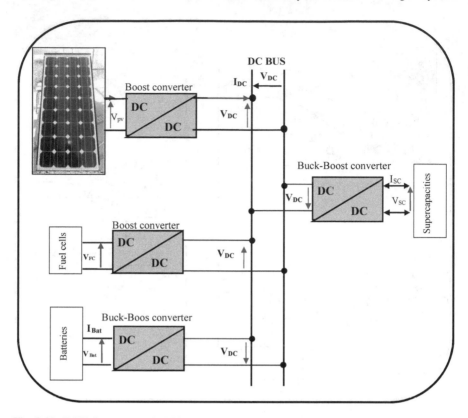

Fig. 2.23 DC/DC converters in PV with multi-storage FCs/batteries

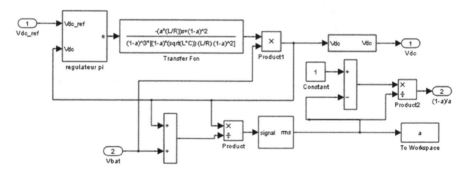

Fig. 2.24 Buck–boost model under MATLAB/Simulink

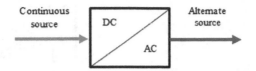

Fig. 2.25 DC/AC converter

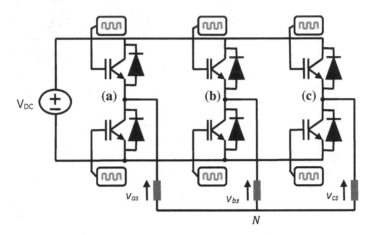

Fig. 2.26 General electrical schematic of three-phase inverter

$$T_{\mathrm{Hi}} = S_i$$
$$T_{\mathrm{Bi}} = \overline{S_i}$$

$$(2.26)$$

where $i = 1, 2, 3$

The electronic scheme for each phase is shown in Figs. (2.27, 2.28, 2.29).

2.3.6.2 Inverter Hysteresis Current Control

The hysteresis control allows an instantaneous adjustment of output current to be maintained below the hysteresis band limited by the upper limit and the lower limit [2], [4] (Fig. 2.30).

The hysteresis control maintains an instantaneous control of the output current under the hysteresis band limited by the upper and lower limits.

Hysteresis current control (HCC) is applied to control the output current of the inverter (Fig. 2.32).

An application under MATLAB/Simulink is given in (Fig. 2.33).

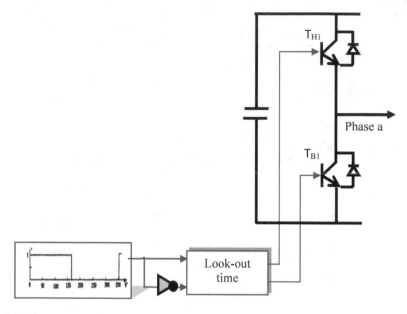

Fig. 2.27 Inverter control

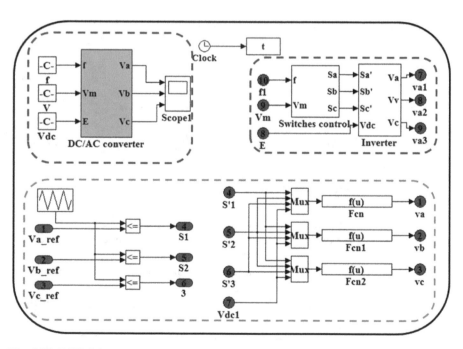

Fig. 2.28 DC/AC inverter model with PWM under MATLAB/Simulink

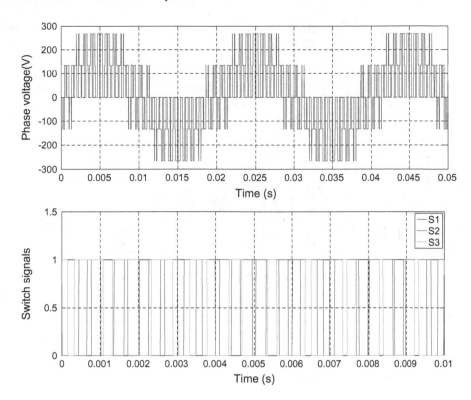

Fig. 2.29 Phase voltage and switch signals in DC/AC converter

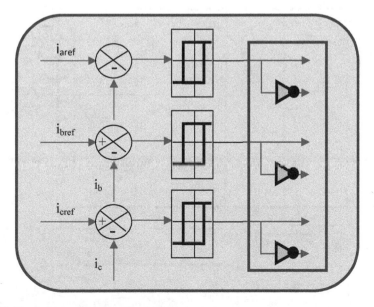

Fig. 2.30 Hysteresis current control principle

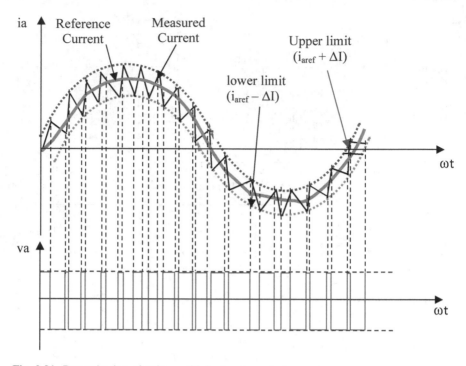

Fig. 2.31 Determination of voltage with hysteresis control current

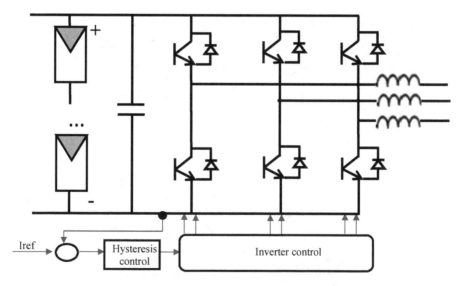

Fig. 2.32 Inverter current control

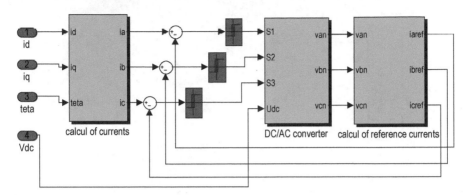

Fig. 2.33 Control hysteresis current under MATLAB/Simulink

2.4 Examples of Applications in PV Systems

In this case, the battery is directly connected to the load (Fig. 2.34).

The DC/DC converter operates as a battery charger and provides energy to a battery bank under controlled voltage and current, in order to improve the service lifetime of the battery bank (Fig. 2.35).

As long as the battery is not fully charged, the control circuit acts on the switches of DC/DC converter in order to maximize the battery current. However, it can estimate the charge of the battery using measurements of battery voltage, current

Fig. 2.34 Simple DC system

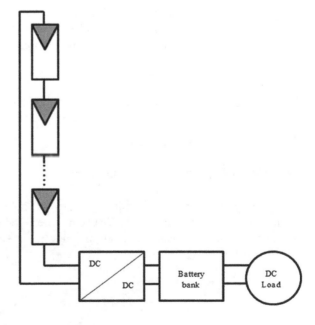

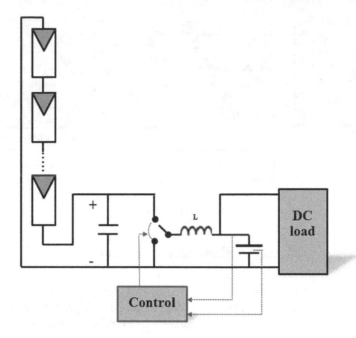

Fig. 2.35 Simplified model of the battery charger

and often temperature and current integral. It can thus limit the battery current if its value is too large considering the battery state of charge. Often, the control circuit can also act on a relay in order to disconnect the load when the battery is too discharged. In theory, a model of the battery is needed in order to estimate the battery state of charge. Several control circuits include a fuzzy logic controller which can determinate implicitly such a model after some cycles of charge.

2.4.1 DC/AC Conversion

2.4.1.1 Converters Topologies

DC/AC Converters (Single Stage Inverters)

When the power generated is transmitted to the public network or used by AC devices, it is necessary to use a DC/AC converter. The most popular is the voltage inverter, whose schematic is given in Fig. 2.36 (single-phase inverter) or at Fig. 2.37 (three-phase inverter).

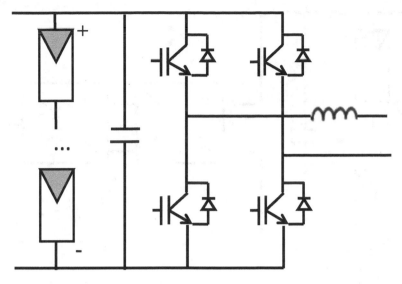

Fig. 2.36 Single-phase voltage inverter

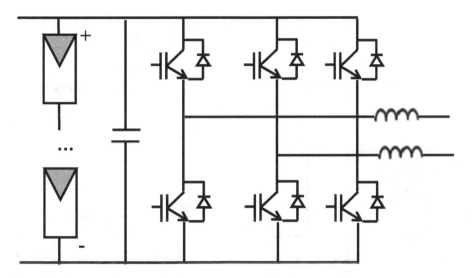

Fig. 2.37 Three-phase voltage inverter

Complex DC/AC Converters (Multistage Inverters)

The drawbacks mentioned in Section "DC/AC Converters (Single Stage Inverters)"
can be avoided adding to the inverter other conversion stages. In fact, what is
named commercially "solar inverter" often includes several power electronic

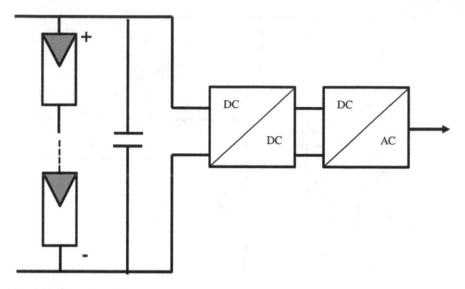

Fig. 2.38 Example of double-stage inverter

converters. As a first example, we can have a dual-stage inverter (Fig. 2.38). This topology consists of a DC/DC converter which is used for the MPPT and a DC/AC inverter.

Another example of double-stage inverter is given in Fig. 2.39.

In Fig. 2.39, the DC link between the DC/DC converter which carries out the MPPT and the inverter is fitted with a capacitor C_{link}.

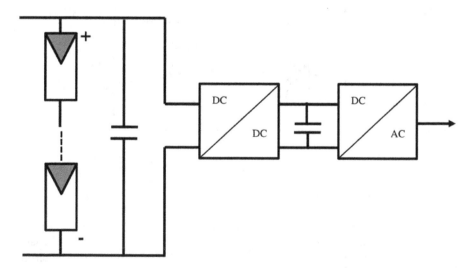

Fig. 2.39 Dual-stage inverter

Multi-input DC/AC Converters

When the optimum working points of the modules are not all the same ones, for example, in the case of partial shade, it can be interesting to split the PV array into several sub-arrays, each of that sub-arrays being fitted with its own converter and MPPT. We can also have a multi-input inverter (Fig. 2.40). This topology consists of two DC/DC converters connected to the DC link of a common DC/AC inverter.

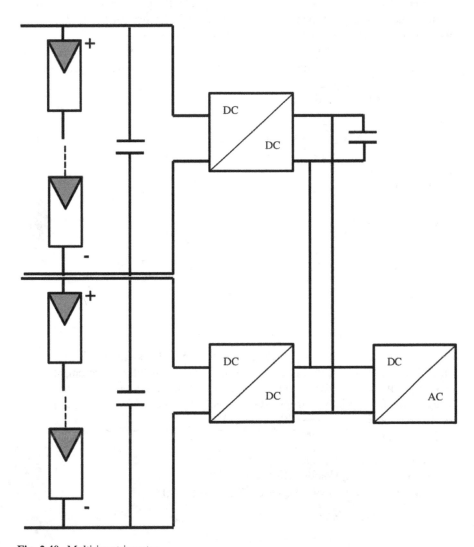

Fig. 2.40 Multi-input inverter

Inverter Architectures

Currently, there are mainly three inverter architectures that provide good technical solutions:

Central Plant Inverter

It is in general used for 10–250 kW PV systems. The PV modules were divided into strings. These ones were then connected in parallel, through string diodes. In this inverter, the DC power supply generated by each string is run along wires to combiner boxes where they are connected in parallel with other strings. From there, the direct current supply is then transferred to the central inverter and converted to alternating current (Fig. 2.41).

The advantages are [26]

- Low cost,
- Fewer component connections,
- Optimal for large systems,
- Low reliability,
- High efficiency,
- Not optimal MPPT.

The disadvantages are:

- Higher installation cost,
- Higher DC wiring and combiner costs.

String Inverters

The string inverter is a reduced version of the centralized inverter. It is used in general for 1.5–5 kW, typical residential application. In these inverters, there are several small inverters for several strings, so that the direct current feed of a few strings runs directly into a string inverter instead of a combiner box and is transformed into alternating current (Fig. 2.42).

The advantages are [26]:

- Lower maintenance costs,
- Simpler design and modularity,
- Each string has its own inverter enabling better MPPT.

The disadvantages are:

- Higher DC watt unit cost,
- More inverter connections.

Fig. 2.41 Central plant inverter

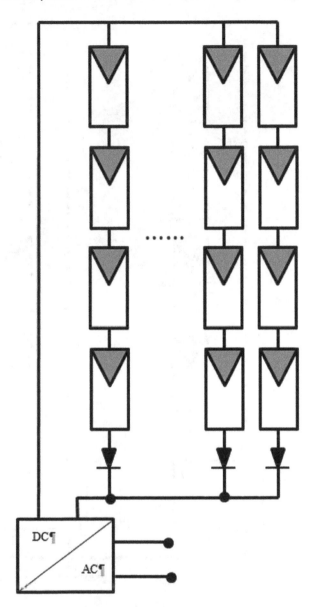

Module Inverters

At the limit case, it is possible to associate a DC/AC converter with each module, which can then be called module inverters (Fig. 2.43). It is used for 50–180 W PV systems. It is an interesting solution when the photovoltaic array is subjected to complex shadows, since each module can always be used at its optimal power. However, the DC/AC converters are complex since there is a large difference in the input and output voltage levels, and their nominal power is low since each of them controls only one module.

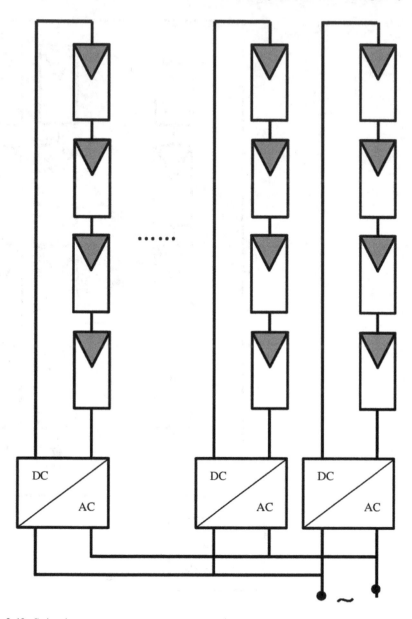

Fig. 2.42 String inverter

The advantages are [26]:

- It has the advantage of being totally modular,
- Simpler design and modularity,
- Each string has its own inverter enabling better MPPT.

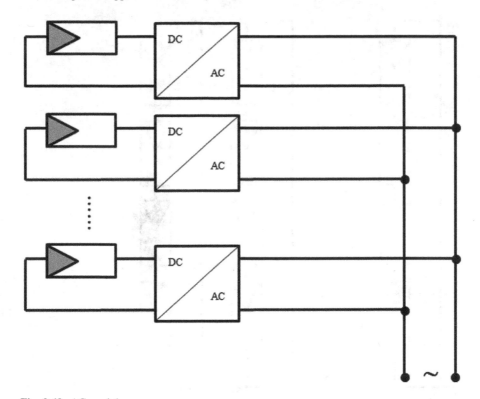

Fig. 2.43 AC modules

The disadvantages are:

- Their efficiency is thus relatively low,
- Difficult maintenance,
- Higher cost/kWp.

Multi-string Inverter

In the multi-string inverter, several strings are interfaced with their own DC/DC converter to a common DC/AC inverter. It is an advantage compared with the centralized system, since every string can be controlled individually. It is also an advantage compared with string inverters and AC modules, since the part of the converter which is critical from the losses point of view is an only DC/AC inverter, which has a higher power and thus a better efficiency (Fig. 2.44).

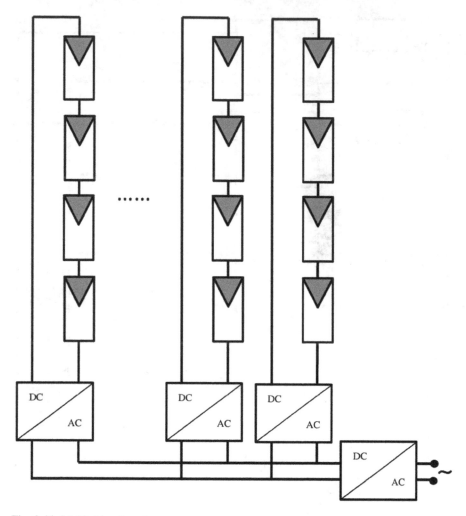

Fig. 2.44 Multi-string inverter

The choice of appropriate configuration must be motivated by the use of conditions imposed by the environment and situation.

2.5 Inverter Control Strategies

2.5.1 Power Control

The inverter is modeled by the equation below [4, 9, 10]:

Fig. 2.45 Direct bus control

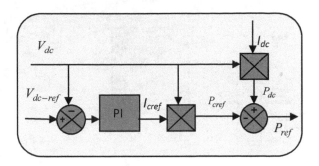

$$\begin{pmatrix} V_{in1} \\ V_{in2} \\ V_{in3} \end{pmatrix} = \frac{V_{dc}}{3} \begin{pmatrix} -2 & 1 & 1 \\ 1 & -2 & 1 \\ 1 & 1 & -2 \end{pmatrix} \begin{pmatrix} F_1 \\ F_2 \\ F_3 \end{pmatrix} \tag{2.28}$$

$V_{in1}, V_{in2}, V_{in3}$ three-phase voltages of the DC/AC inverter. Then the filter, constituted of R_n and L_n, links these voltages to E_1, E_2, E_3, which are the three-phase voltages of the grid, through the following relation:

$$\begin{pmatrix} V_{in1} \\ V_{in2} \\ V_{in3} \end{pmatrix} = R_n \begin{pmatrix} i_{n1} \\ i_{n2} \\ i_{n3} \end{pmatrix} + L_n \frac{d}{dt} \begin{pmatrix} i_{n1} \\ i_{n2} \\ i_{n3} \end{pmatrix} + \begin{pmatrix} E_1 \\ E_2 \\ E_3 \end{pmatrix} \tag{2.29}$$

The control algorithm to find out the reference active is presented in Fig. 2.45.

The network link control consists in adjusting the active power supplied to the grid to its reference value and the reactive power to zero in order to fix the power factor at the unit. The active power reference is deduced by controlling the direct bus voltage with a proportional integral corrector generating the current reference to the capacity. Hence, we can express P_{ref} as (Fig. 2.31):

$$P_{ref} = V_{dc} \cdot (I_{dc} - I_{cref}) \tag{2.30}$$

$$P_{ref} = P_{dc} - P_{cref} \tag{2.31}$$

$$i_{cref} = PI. (V_{dc-ref} - V_{dc}) \tag{2.32}$$

The reference active and reactive powers are given by the following equations (Fig. 2.31):

$$P_{ref} = E_d \cdot i_{nd\_ref} + E_q \cdot i_{nq\_ref} \tag{2.33}$$

$$Q_{ref} = E_q \cdot i_{nd\_ref} - E_d \cdot i_{nq\_ref} \tag{2.34}$$

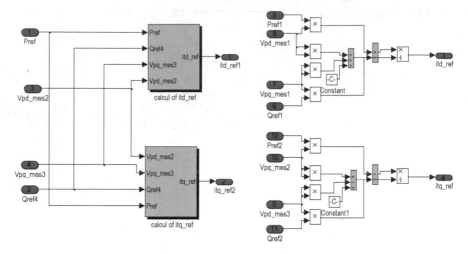

Fig. 2.46 Calcul of current references

Then multiplying Eq. 2.33 by E_d and Eq. 2.34 by E_d, the addition and subtraction of the two resulting equations give the reference current value according to active and reactive power ones by [27] (Fig. 2.46):

$$i_{nd-ref} = \frac{P_{ref}E_d + Q_{ref}E_q}{E_d^2 + E_q^2} \qquad (2.35)$$

$$i_{nq-ref} = \frac{P_{ref}E_q - Q_{ref}E_d}{V_d^2 + V_q^2} \qquad (2.36)$$

Then the global control system is described in Figs. 2.47, 2.49.

2.6 Applications to Pumping Systems

In solar pumping systems, the system architecture with the different converter can be represented as (Fig. 2.48) [3, 16].

2.7 Application in Wind Turbine Energy Conversion

In wind turbine systems, the DC/DC converter can be inserted between the AC/DC and the DC/AC converter [28] (Fig. 2.50).

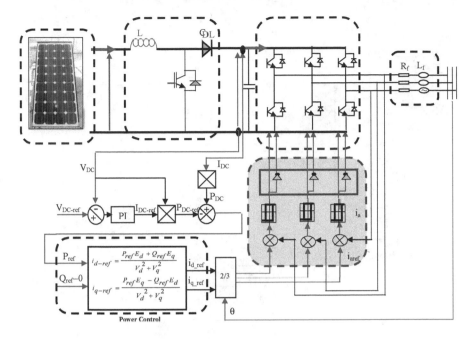

Fig. 2.47   Power control of PV system connected to the grid

2.7.1   AC/DC Converters

A rectifier or an AC/DC converter consists in converting an alternating voltage or current into its direct voltage or current. It can be a controlled rectifier (Fig. 2.51) or uncontrolled rectifier which consists of bridge diodes (Figs. 2.52, 2.53).

2.7.2   AC/AC Converters

The problem with photovoltaic systems is how to integrate it into the grid using power converters. Usually, it generates harmonics in the network. To address this constraint, it is necessary to use a multilevel inverter. Indeed, the various studies on multilevel inverters have shown that they offer a great improvement on the quality of output voltages. To further improve these voltages, different strategies controls have been used. These strategies allow us to eliminate some ranks of harmonics (Fig. 2.54) [29].

The structure of PV system connected to the grid using multi-cell inverter and controlled by FOC strategy is shown in Fig. 2.55.

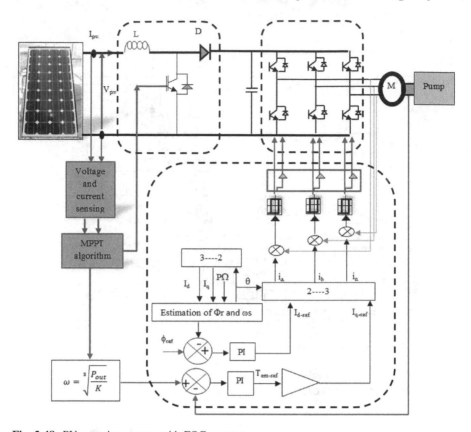

Fig. 2.48 PV pumping system with FOC strategy

2.8 Examples of Converter Topologies in Hybrid Systems

2.8.1 Hybrid PV/DG/Wind Turbine with Battery Storage

In this topology, wind turbine, diesel generator and photovoltaic sources are incorporated together using a combination of DC/DC and DC/AC converters (Fig. 2.56).

2.8.2 Hybrid PV/Wind System

In this configuration, both wind and photovoltaic sources are incorporated together using a combination of boost and buck–boost converters (Fig. 2.57).

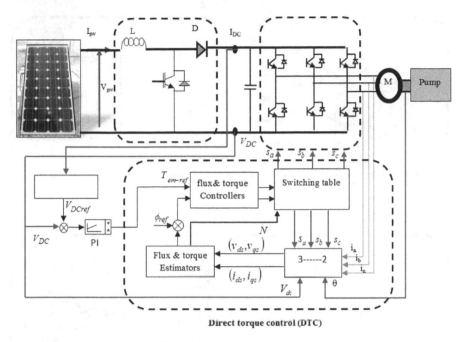

Direct torque control (DTC)

Fig. 2.49 PV pumping system with DTC strategy

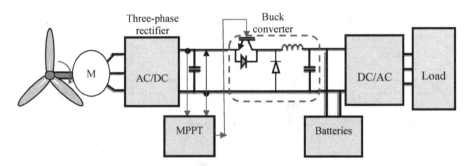

Fig. 2.50 DC/DC converter in wind turbine system

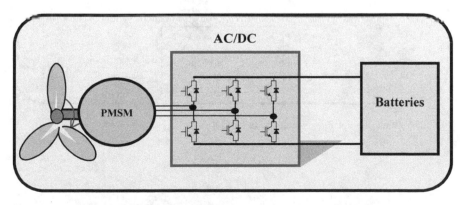

Fig. 2.51 Wind system with controlled rectifier

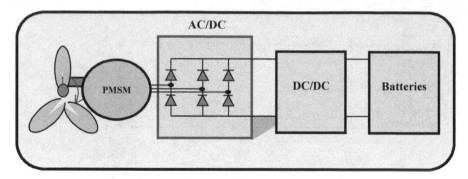

Fig. 2.52 Uncontrolled rectifier

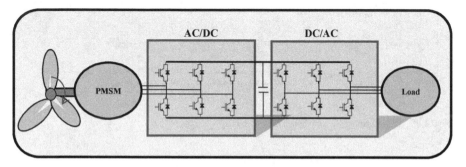

Fig. 2.53 The back-to-back PWM-VSI converter in wind system

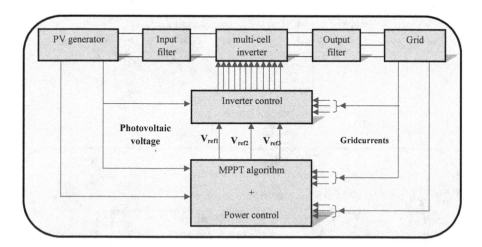

Fig. 2.54 Connection of PV generator to the grid

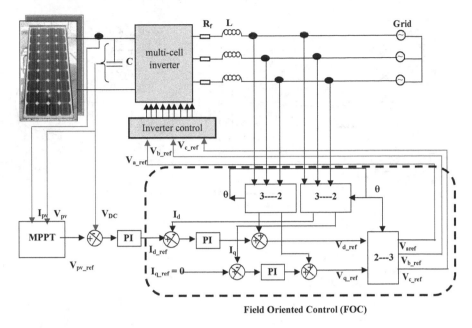

Fig. 2.55 PV system using multi-cell inverter and controlled with FOC method

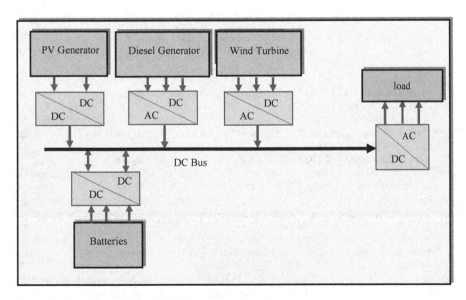

Fig. 2.56 Used converters in PV/DG/wind system with battery storage

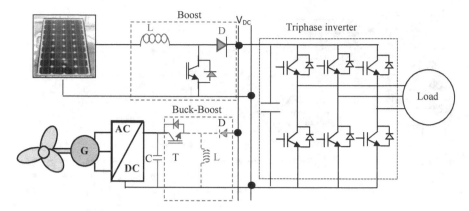

Fig. 2.57 Used converters in PV/wind system

References

1. Jha M, Blaabjerg F, Khan MA, Kurukuru VSB, Haque A (2019) Intelligent control of converter for electric vehicles charging station. Energies 12(12), art. no. 2334
2. Rekioua D, Rekioua T (2005) A new approach to direct torque control strategy with minimization torque pulsations in permanent magnets synchronous machines. In: 2005 IEEE Russia Power Tech, Power Tech, art. no. 4524699
3. Serir C, Rekioua D (2015) Control of photovoltaic water pumping system. J Electr Eng 15 (2):339–344
4. Rekioua D, Rekioua T, Soufi Y (2015) Control of a grid connected photovoltaic system. In: 2015 international conference on renewable energy research and applications (ICRERA 2015), pp 1382–1387, art. no. 7418634
5. Zhang Z, Zhang Z, Cao Y, Wu Z, Qian Q, Xie S (2018) Research on two current-fed isolated bidirectional DC/DC converters for the battery energy storage application. In: Proceedings— 2017 IEEE southern Power Electronics Conference (SPEC 2017), Jan 2018, pp 1–6
6. Aissou R, Rekioua T, Rekioua D, Tounzi A (2016) Robust nonlinear predictive control of permanent magnet synchronous generator turbine using Dspace hardware. Int J Hydrogen Energy 41(45):21047–21056
7. Rekioua D, Rekioua T (2009) DSP-controlled direct torque control of induction machines based on modulated hysteresis control In: Proceedings of the International Conference on Microelectronics (ICM), pp 378–381, art. no. 5418603
8. Rekioua T, Rekioua D (2003) Direct torque control strategy of permanent magnet synchronous machines. In: 2003 IEEE Bologna power tech—conference proceedings, vol 2. pp. 861–866, art. no. 1304660
9. Abdelli R, Rekioua D, Rekioua T, Tounzi A (2013) Improved direct torque control of an induction generator used in a wind conversion system connected to the grid. ISA Trans 52 (4):525–538
10. Abdelli R, Rekioua D, Rekioua T (2011) Performances improvements and torque ripple minimization for VSI fed induction machine with direct control torque. ISA Trans 50(2):213–219
11. Benyahia N, Denoun H, Badji A, Zaouia M, Rekioua T, Benamrouche N, Rekioua D (2014) MPPT controller for an interleaved boost DC–DC converter used in fuel cell electric vehicles. Int J Hydrogen Energy 39(27):15196–15205

12. Wang H, Gaillard A, Hissel D (2019) A review of DC/DC converter-based electrochemical impedance spectroscopy for fuel cell electric vehicles. Renew Energy 124–138

13. Azri M, Khanipah NHA, Ibrahim Z, Rahim NA (2017) Fuel cell emulator with MPPT technique and boost converter. Int J Power Electron Drive Syst 8(4):1852–1862

14. Samal S, Ramana M, Barik PK (2018) Modeling and simulation of interleaved boost converter with MPPT for fuel cell application. In: 2018—proceedings international conference on technologies for smart city energy security and power: smart solutions for smart cities (ICSESP), Jan 2018, pp 1–5

15. Iqbal M, Benmouna A, Eltoumi F, Claude F, Becherif M, Ramadan HS (2019) Cooperative operation of parallel connected boost converters for low voltage-high power applications: an experimental approach. Energy Procedia 162:349–358

16. Mohammedi A, Rekioua D, Mezzai N (2013) Experimental study of a PV water pumping system. J Electr Syst 9(2):212–222

17. Ravi D, Letha SS, Samuel P, Reddy BM (2018) An overview of various DC–DC converter techniques used for fuel cell based applications. In: International conference on power energy, environment and intelligent control (PEEIC), pp 16–21, art. no. 8665465

18. Dixon RC, Mikhalchenko GYa, Mikhalchenko SG, Russkin VA, Semenov SM (2017) Issues of linearization of a two-phase boost DC–DC converter applied in the power supply systems operating on renewable energy sources. Bull Tomsk Polytechnic Univ Geo Assets Eng 328 (1):89–99

19. Osipov AV, Zapolskiy SA (2018) Boost type resonant lcljt converter for autonomous power supply system from renewable sources. Bull Tomsk Polytechnic Univ Geo Assets Eng 329 (3):77–88

20. Abu-Aisheh AA (2019) Design and analysis of solar/wind power electronics converters. Renew Energy Power Qual J 17:349–353

21. Ghiasi M (2019) Detailed study, multi-objective optimization, and design of an AC–DC smart microgrid with hybrid renewable energy resources. Energy 169:496–507

22. Aissou S, Rekioua D, Rekioua T, Bacha S (2019) Simple and low-cost solution system for a small scale power photovoltaic water pumping system. In: Proceedings of 2018 6th international renewable and sustainable energy conference (IRSEC), art. no. 8702980

23. Zeng J, Ning J, Kim T, Winstead V (2019) Modeling and control of a four-port DC–DC converter for a hybrid energy system. In: Conference proceedings—IEEE applied power electronics conference and exposition–APEC, Mar 2019, pp 193–198, art. no. 8722323

24. Ferchichi M, Zaidi N, Khedher A (2016) Comparative analysis for various control strategies based MPPT technique of photovoltaic system using DC–DC boost converter. In: Proceedings 2016 17th international conference on Sciences and Techniques of Automatic control and Computer Engineering (STA), pp. 532–539, art. no. 7951990, ISBN: 978-150903407-9. https://doi.org/10.1109/sta.2016.7951990

25. Blaabjerg F, Iov F, Teodorescu R, Chen Z (2006) Power electronics in renewable energy systems. In: 12th international power electronics and motion control conference, Portoroz, Slovenia, 30 Aug–1 Sept 2006

26. Blaabjerg F, Chen Z (2003) Power electronics as an enabling technology for renewable energy integration. J Power Electr 3(2):81–89

27. Maheri A (2016) Effect of dispatch strategy on the performance of hybrid wind-PV battery-diesel-fuel cell systems. J Therm Eng 2(4):820–825

28. Rekioua D 2014 Wind power electric systems: modeling, simulation and control. In: Green energy and technology. Springer, Heidelberg

29. Rahrah K, Rekioua D, Rekioua T (2015) Optimization of a photovoltaic pumping system in Bejaia (Algeria) climate. J Electr Eng 15(2):321–326

Chapter 3
MPPT Methods in Hybrid Renewable Energy Systems

3.1 Introduction to Optimization Algorithms

The most used control technique in optimization consists in acting on the duty cycle automatically to place the generator (PV, wind turbine, etc.) at its optimal value whatever the variations on the metrological conditions or sudden changes in loads which can occur at any time. In general, there are two types of algorithms: classical and advanced [1–59]. In this chapter, the most used MPPT methods in PV and wind turbine systems are presented. For each method, it is given the concept or the principle, the algorithm, the flowchart, blocks schemes under MATLAB/Simulink and for some methods an application with obtained results in MATLAB. All these details help the user to better understand these algorithms.

3.2 MPPT for Hybrid System

In a hybrid system (PV, wind, hydropower, etc.), the power is maximized using MPPT controllers. An example of PV/wind turbine/FCs system with MPPT controllers is given (Fig. 3.1).

3.3 Survey of Maximum Power Point Tracking (MPPT) Algorithms in PV Systems

The output power induced in the photovoltaic modules depends on solar irradiance and temperature of the solar cells. Therefore, to maximize the efficiency of the renewable energy system, it is necessary to track the maximum power point of the PV array. The PV array has a unique operating point that can supply maximum

© Springer Nature Switzerland AG 2020
D. Rekioua, *Hybrid Renewable Energy Systems*, Green Energy and Technology,
https://doi.org/10.1007/978-3-030-34021-6_3

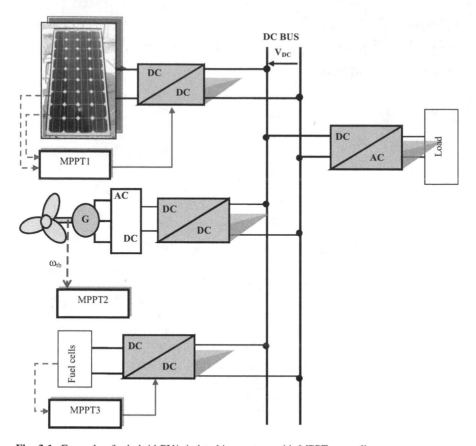

Fig. 3.1 Example of a hybrid PV/wind turbine system with MPPT controllers

power to the load. This point is called the maximum power point (MPP). The locus of this point has a nonlinear variation with solar irradiance and the cell temperature [60]. Thus, in order to operate the PV array at its MPP, the PV system must contain a maximum power point tracking (MPPT) controller (Fig. 3.2).

The maximum power point (MPP) is obtained when the derivative of PV power by the voltage ($\frac{\Delta P_{pv}}{\Delta V_{pv}}$) is zero. Basically, to achieve the maximum power point of operation, the generator voltage V_{pv} is regulated so that it increases when the slope $\frac{\Delta P_{pv}}{\Delta V_{pv}}$ is positive and it decreases when the slope $\frac{\Delta P_{pv}}{\Delta V_{pv}}$ is negative. A control which provides continuous extraction of maximum power point is given by:

$$V_{opt} = K_G . \int \frac{dP_{pv}}{dV_{pv}} \, dt \approx K_G . \int \frac{\Delta P_{pv}}{\Delta V_{pv}} \, dt \tag{3.1}$$

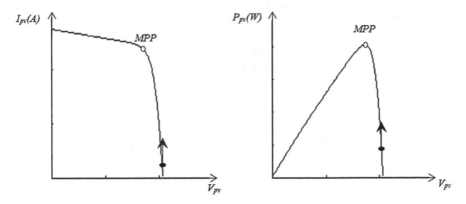

Fig. 3.2 Photovoltaic electrical characteristics

where V_{opt} is the optimal voltage which gives maximum power, K_G is a proportional gain, ΔP_{pv} is power variation between two operating points and ΔV_{pv} is voltage variation between two operating points. Many algorithms have been developed to determine the maximum power point (MPP) [48].

3.3.1 Most Used MPPT Algorithms in Photovoltaic Systems

3.3.1.1 Perturb and Observe (P&O) Technique

P&O is one of the most used algorithms. This method is very simple to be implemented and it does not require knowledge of the photovoltaic parameters [32]. Figure 3.3 shows the variation of the duty ratio or voltage depending on the power.

The algorithm is as follows:

When $\Delta P_{pv} \rangle 0$ and $\Delta V_{pv} \rangle 0$ thus $\frac{\Delta P_{pv}}{\Delta V_{pv}} \rangle 0$, we increment the duty cycle D, $D = D + \Delta D$.

When $\Delta P_{pv} \langle 0$ and $\Delta V_{pv} \langle 0$ thus $\frac{\Delta P_{pv}}{\Delta V_{pv}} \langle 0$, we decrement the duty cycle, $D = D - \Delta D$.

When $\Delta P_{pv} = 0$ and $\Delta V_{pv} = 0$ thus $\frac{\Delta P_{pv}}{\Delta V_{pv}} = 0$, we retain the duty cycle D, $\Delta D = 0$ and $D = D$.

The flowchart of the P&O algorithm is given in Fig. 3.4.
The implementation under MATLAB/Simulink can be represented in Fig. 3.5.

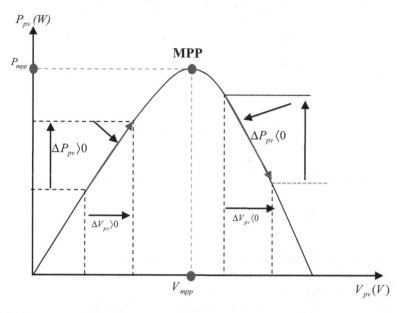

Fig. 3.3 Variation of the voltage depending on the power

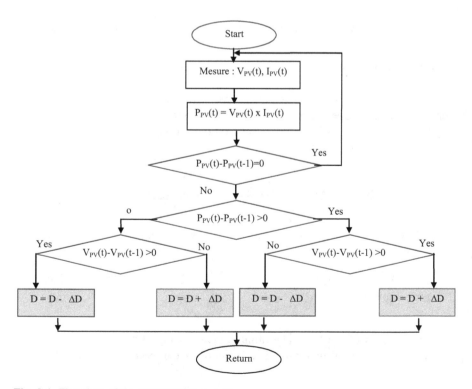

Fig. 3.4 Flowchart of the P&O MPPT algorithm

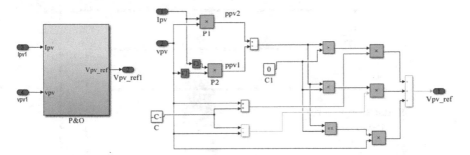

Fig. 3.5 P&O algorithm under MATLAB/Simulink

An example of maximum power point tracking code for a photovoltaic panel is given in Fig. 3.6.

The implementation under MATLAB/Simulink is in Fig. 3.7.

Some simulation results are given in Fig. 3.8.

3.3.1.2 Modified P&O Method (MP&O)

An improved P&O algorithm is proposed to overcome the disadvantage of the P&O method especially under weather changes (Fig. 3.9).

Sixteen possibilities can be possible (Table 3.1). We note that when increasing the panel power on two disturbances in the same direction, there is a new condition which is introduced.

The implementation under MATLAB/Simulink can be represented in Fig. 3.10.

3.3.1.3 Incremental Conductance Algorithm (IC or IncCond)

This method focuses directly on power variations. The output current and voltage of the photovoltaic panel are used to calculate the conductance and the incremental conductance. Its principle is to compare the conductance [61] (cond $= \frac{I_{pv}}{V_{pv}}$) and incremental conductance (dcond $= \frac{dI_{pv}}{dV_{pv}}$) and to decide when to increase or to decrease the PV voltage to reach the MPP where the derivative of the power is equal to zero ($\frac{dP_{pv}}{dV_{pv}} = 0$) (Fig. 3.11). The disadvantage of this method with fixed step is its oscillations [17, 50].

The algorithm is in Fig. 3.12.

The flowchart of the Incremental algorithm is given in Fig. 3.13.

The implementation under MATLAB/Simulink can be represented in Fig. 3.14.

Some simulation results are given under sudden variation of solar radiation (Fig. 3.15).

```
function y=mpp(u)
v1=v(1);
  I1=I(1);
for i=1:29
  P(i)=v(i)*I(i);
   if  v1*I1<v(i)*I(i)
      v1=v(i);
      I1=I(i);
   else
      v1=v1;
   I1=I1;
end
end
v1;
I1;
P=v1*I1;
If=P/24;
vf= 24;
y1=v1;
y2=I1;
y3=If;
y4=vf;
y= [y1 y2 y3 y4].
```

Fig. 3.6 Maximum power point tracking code for a photovoltaic panel

3.3.1.4   Constant Reference Voltage Algorithm

The simplest technique to maintain the operation of the PV system around the maximum power point is to control the voltage measured at the PV generator, to its reference voltage corresponding to the optimum voltage [62]. This method assumes that the variation of the optimum voltage to climatic factors variations (insolation, temperature) is negligible as shown in Fig. 3.16a. However, when the junction temperature of the PV cell varies as shown in Fig. 3.16b, the optimum voltage will be not constant. This method uses a single control loop and is well suited for applications where climatic conditions are stable, such as space satellites [63].

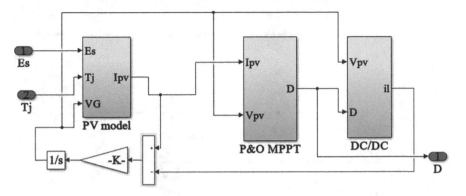

Fig. 3.7 PV system with P&O MPPT

3.3.1.5 Open-Circuit Voltage (OCV) Algorithm

This technique is very simple. It consists in comparing the photovoltaic voltage with a reference voltage corresponding to the optimal voltage. The voltage error is then used to adjust the duty cycle of the converter to coincide the two voltages. The reference voltage is obtained from the knowledge of the existing linear relationship between the optimal voltage and the open-circuit voltage [48].

$$V_{opt} = K_v.V_{oc} \tag{3.2}$$

From Eq. (3.2) and with knowledge of K_v, it is possible to measure the open-circuit voltage (V_{oc}) in order to deduce the reference voltage (V_{opt}) to be applied to the PV as shown in Fig. 3.17.

3.3.1.6 Fractional Short-Circuit Current (SSCC) Algorithm

This method is based on the knowledge of the linear relationship between the optimal current (I_{opt}) and the short-circuit current (I_{cc}) (Fig. 3.18).

$$I_{opt} = K_c.I_{sc} \tag{3.3}$$

where k_c is a current factor that also depends on the GPV used and is generally between 0.78 and 0.92 [48].

The disadvantage of this method is the complexity of measuring the short-circuit current I_{sc} while the system is in operation.

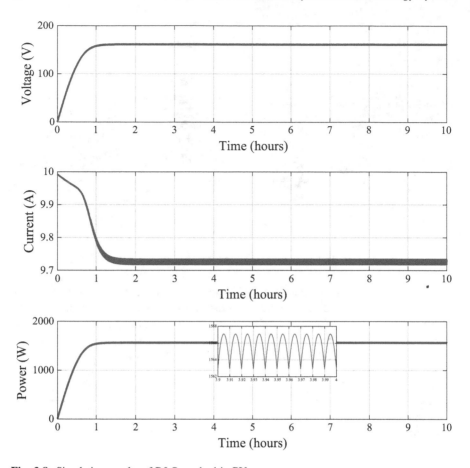

Fig. 3.8 Simulation results of P&O method in PV system

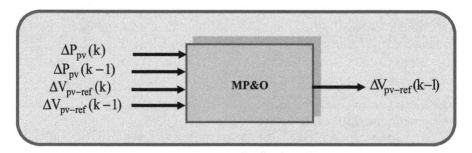

Fig. 3.9 Modified P&O model

Table 3.1 Rule table of modified P&O

Input variables				Output variable	System state
$\Delta V_{pv-ref}(k-1)$	$\Delta P_{pv}(k-1)$	$\Delta V_{pv}(k)$	$\Delta P_{pv}(k)$	$\Delta V_{pv-ref}(k-1)$	
<0	<0	<0	<0	>0	Invalid
<0	<0	<0	$>$	>0	Invalid
<0	<0	$>$	<0	<0	Decreasing of G
<0	<0	>0	>0	>0	$V_{pv} < V_{mpp}$
<0	>0	<0	<0	>0	$V_{pv} \approx V_{mpp}$
<0	>0	<0	>0	>0	New condition
<0	>0	>0	<0	<0	$V_{pv} > V_{mpp}$
<0	>0	>0	>0	<0	Increasing of G
>0	<0	<0	<0	>0	Decreasing of G
>0	<0	<0	>0	<0	$V_{pv} > V_{mpp}$
>0	<0	>0	<0	<0	Invalid
>0	<0	>0	>0	<0	Invalid
>0	>0	<0	<0	>0	$V_{pv} < V_{mpp}$
>0	>0	<0	>0	>0	Increasing of G
>0	>0	>0	<0	<0	$V_{pv} \approx V_{mpp}$
>0	>0	>0	>0	<0	New condition

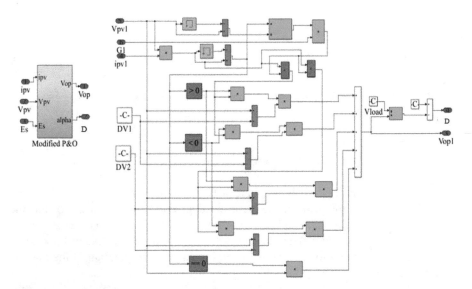

Fig. 3.10 Modified P&O under MATLAB/Simulink

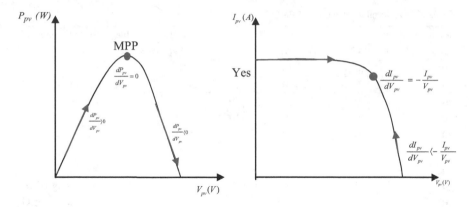

Fig. 3.11   Incremental conductance algorithm principle

When $\dfrac{dI_{pv}}{dV_{pv}} > -\dfrac{I_{pv}}{V_{pv}}$, the duty cycle is incremented, D $D = D + \Delta D$

When $\dfrac{dI_{pv}}{dV_{pv}} < -\dfrac{I_{pv}}{V_{pv}}$, the duty cycle is decremented D, $D = D - \Delta D$

When $\dfrac{dI_{pv}}{dV_{pv}} = -\dfrac{I_{pv}}{V_{pv}}$, the duty cycle is retained D, $\Delta D = 0$, $D = D$

Fig. 3.12   IncCond algorithm

3.3.1.7   Look-Up Table Algorithm

It consists of measuring the photovoltaic voltage and current and comparing them with the data in memory under the same weather conditions in order to find the maximum power point (MPP). The disadvantage of this method is the need for a large storage memory; moreover, the algorithm may prove ineffective in some cases since it is very difficult to store all the characteristics corresponding to the various weather situations [48].

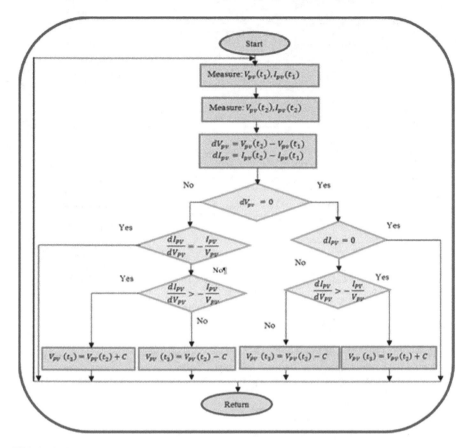

Fig. 3.13 Flowchart of the Inc MPPT algorithm

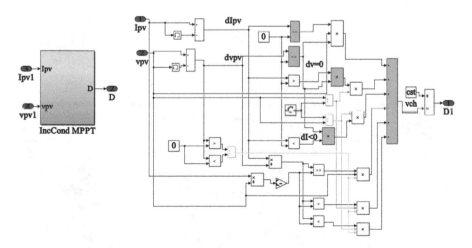

Fig. 3.14 IncCond MPPT method under MATLAB/Simulink

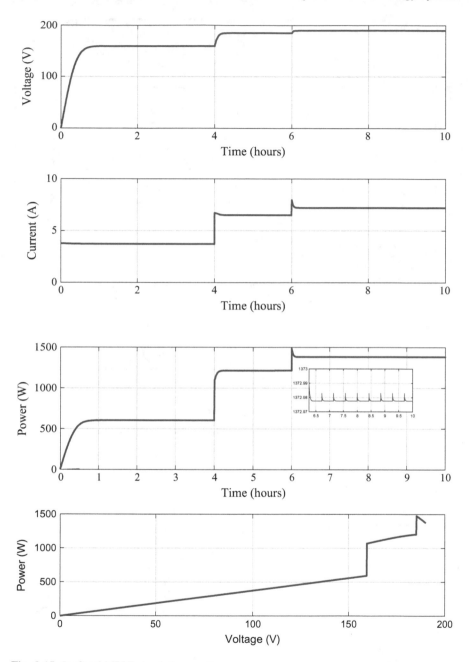

Fig. 3.15 IncCond MPPT simulation results

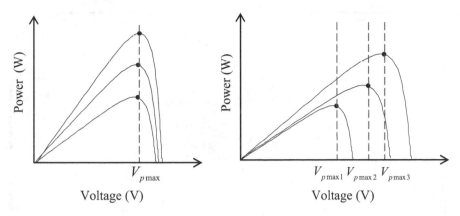

Fig. 3.16 Characteristic power-voltage $P_{pv}(V_{pv})$ of PV generator

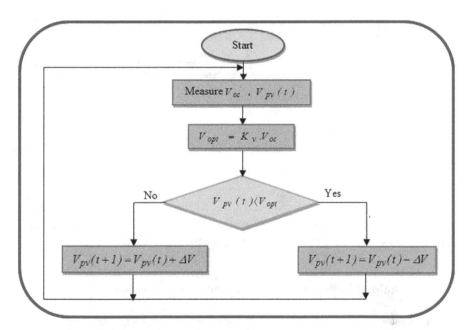

Fig. 3.17 Flowchart of the OCV MPPT algorithm

3.3.1.8 Curve-Fitting Method

It is based on the exact knowledge of the characteristic of photovoltaic panels from which mathematical equations are extracted to determine the maximum power point.

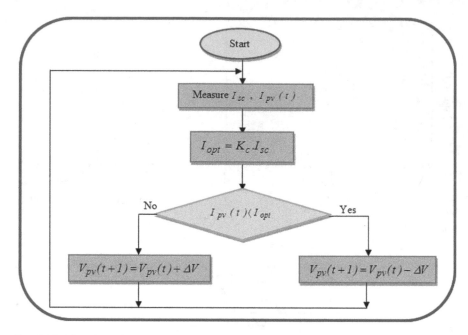

Fig. 3.18 Flowchart of the SC MPPT algorithm

$$V_{pv} = a..V_{pv}^3 + b..V_{pv}^2 + c..V_{pv} + d \tag{3.4}$$

$$V_{opt} = \frac{-b\sqrt{b^2 - 3ac}}{3a} \tag{3.5}$$

The main disadvantages of this method are the high number of iterations required to obtain V_{opt} and the need for large memory capacity.

3.3.1.9 Optimal Voltage with Temperature Compensation

We generate the reference voltage by adding a unique cell junction which is electrically independent of the PV system and having an electrical characteristic identical to the cells of the PV generator (Fig. 3.19).

The open-circuit voltage V_{oc} varies with the cell temperature T_j, and the short-circuit current I_{sc} is directly proportional to the irradiance level G. It can be described through the following equation:

$$V_{oc} = V_{oc-STC} + \gamma(T_j - T_{j-STC}) \tag{3.6}$$

where V_{oc-STC} is the open-circuit voltage under standard test conditions (V), γ is the temperature coefficient (V/K) and T_{j-STC} is the cell temperature under STC (K).

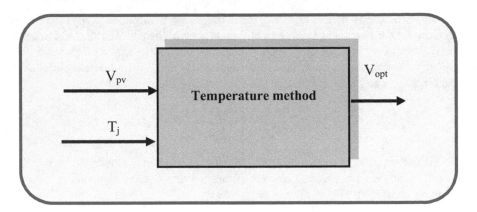

Fig. 3.19 Block diagram of temperature method

3.3.1.10   Method of a Model Parasitic Capacity (PC)

The algorithm of the parasitic capacitance is similar to the increment of the conductivity except that the effect of parasitic capacitance (C_P) which models the storage charges in the junction's p-n solar cells is included. By adding this capability to our model by representing it as:

$$I_{pv} = C_P \frac{dV_{pv}}{dt} \tag{3.7}$$

The new model can be expressed as [64, 65]:

$$
\begin{aligned}
I_{pv} &= I_{ph} - I_S \left[\exp\left(\frac{(V_{pv} + R_s I_{pv})}{A V_{th}} \right) - 1 \right] - \left(\frac{V_{pv} + R_s I_{pv}}{R_{Sh}} \right) + C_P \frac{dV_{pv}}{dt} \\
&= F(V_{pv}) + C_P \frac{dV_{pv}}{dt}
\end{aligned}
\tag{3.8}
$$

Equation (3.8) shows the two components (I_{pv}) is a function of the voltage F (V) and the second relates to the current in the stray capacitance. Using this notation, the increment of the conductivity of the solar panel can be defined as the ratio $dF(V_{pv})/dV_{pv}$ and the instantaneous conductivity can be defined as the ratio $F(V_{pv})/V_{pv}$. The MPP is obtained when $dP_{pv}/dV_{pv} = 0$.

Multiplying Eq. (3.8) by the voltage (V_{pv}) panel for electric power, and then differentiating the result, the equation of electric power at MPP is obtained and can be expressed as [66]:

$$\frac{dF(V_{pv})}{dV_{pv}} + C_{pv}\left(\frac{\dot{V}_{pv}}{V_{pv}} + \frac{\ddot{V}_{pv}}{\dot{V}_{pv}} \right) + \frac{F(V_{pv})}{V_{pv}} = 0 \tag{3.9}$$

The three terms of Eq. (3.9) represent the increase in conductivity, the wave induced by the parasitic capacitance and conductivity instant. The first and second derivatives of the voltage of the panel take into account the ripple effect generated by the alternative converter. Note that if (C_{Pv}) is zero, Eq. (3.9) simplifies and becomes the one used for the algorithm to increase the conductivity.

$$\frac{dF(V_{pv})}{dV_{pv}} + \frac{F(V_{pv})}{V_{pv}} = 0 \tag{3.10}$$

Since the parasitic capacitance is modeled as a capacitor connected in parallel across each cell photovoltaic panels connected in parallel to increase overall capacity for the MPPT.

3.3.1.11 Switching Ripple Correlation Control (SRCC) Method

In switching ripple correlation control (SRCC), ripple in PV voltage and current are used to perform MPPT. The switching frequency of the converter controls the time convergence. This method is simple, applicable with analog circuits and tracks MPP quickly and we have no dependency of PV array.

The ripple of current and voltage is used to obtain maximum power tracking. The duty ratio can be written as:

$$D(t) = -k. \int \dot{p}.\dot{v} \,.dt \tag{3.11}$$

3.3.1.12 Particle Swarm Optimization (PSO) Method

A swarm is a population of particles and each particle flies toward the optimum or a quasi-optimum solution based on its own experience, experience of nearby particles and global best position among particles in the swarm [1, 40, 45]. At time t, each particle i has its position X_t^i and velocity V_t^i in a variable space (Fig. 3.20). The velocity and position of each particle change in the next generation (X_i^{k+1} and V_i^{k+1}).

Velocity and position update formula are as follows [51–53]:

$$V_i^{k+1} = wV_i^k + c_1 r_1 (P_i^k - X_i^k) + c_2 r_2 (P_g^k - X_i^k) \tag{3.12}$$

$$X_i^{k+1} = X_i^k + V_i^{k+1} \tag{3.13}$$

where i individual particle, V_i^{k+1} is the particle velocity, X_i^{k+1} is the current position of a particle, P_i^k is the Pbest and P_g^k is the Gbest, r_1 and r_2 is the random number between 0 and 1 and c_1 and c_2 are learning factors (usually $c_1 = c_2 = 2$).

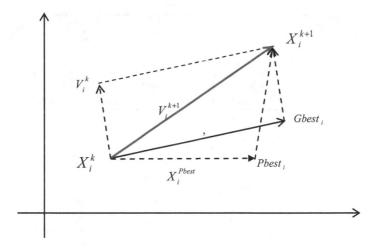

Fig. 3.20 PSO concept

The algorithm is in Fig. 3.21.

And the PSO flowchart is in Fig. 3.22.

Application under MATLAB/Simulink: Particle swarm optimization simulation parameters ($c_1 = 1$; $c_2 = 4 - c_1$; $C = 1$; max iteration = 50: number of particles = 10). Some simulation results are given in Figs. 3.23, 3.24 and 3.25.

3.3.1.13   Ant Colony Optimization (ACO) Method

Metaheuristics are a set of methods used to solve optimization problems that are considered difficult. The objective function of the ACO is to minimize the error function between the speed reference and the machine speed [6]. The ant colony algorithm is in Fig. 3.26.

The algorithm is developed in Fig. 3.27.

> ➢ **Step 1-** Set the number of particles and searching parameters along with the limit for position and velocity.
> ➢ **Step 2-** Randomly initialize Position and velocity of each particle.
> ➢ **Step 3-** Compute the fitness value of each particle.
> ➢ **Step 4-** The particle having the best fitness value is set as Gbest (Global Best).
> ➢ **Step 5-** Update the position and velocity of each particle with respect to the Gbest.
> ➢ **Step 6-** Repeat Step 3 & 4 till the optimum solution is reached.
> ➢ **Step 7-** Gbest at the end of the last iteration gives the optimized value.
> ➢ **Step 8-** Compute the Duty-cycle. $D = \dfrac{1}{1 + \sqrt{\frac{R_{in}}{R_{out}}}}$

Fig. 3.21 PSO algorithm

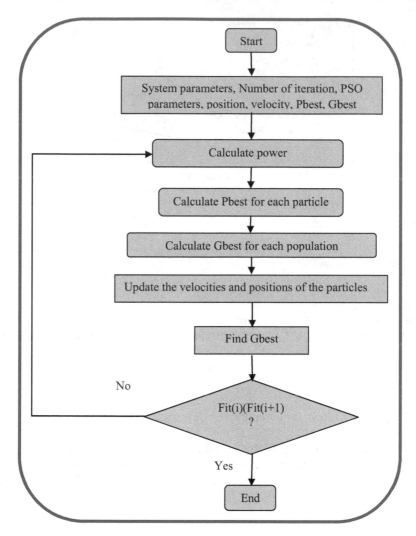

Fig. 3.22 PSO flowchart

The metaheuristics-based optimized algorithm has been invented by Amer et al. [7] to solve difficult nonlinear issues. The particular ant to obtain the shortest path optimization generates pheromones. For food hunting, the movements of ants take place in different direction followed with generated pheromones. The shortest path should have high pheromones probability as it evaporates in short and methodology is repeated for different iterations to optimize the problems.

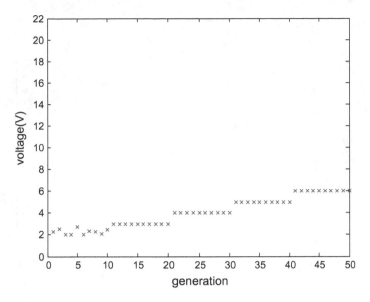

Fig. 3.23 Voltage particles motion

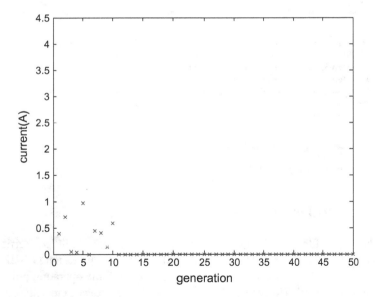

Fig. 3.24 Current particles motion

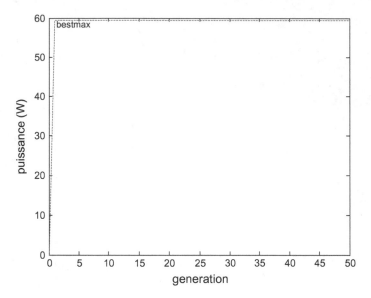

Fig. 3.25 Objective function of particles

> ➤ **Step 1:** Create a population of "N" ants, initialize the pheromone and parameters.
> ➤ **Step 2:** For each individual ant, evaluate the fitness function (ITAE).
> ➤ **Step 3:** Determine the best solution for each ant;
> ➤ **Step 4:** Update the pheromone.
> ➤ **Step 5:** Check if the convergence is satisfied (all ants must have the same solution value);
> ➤ **Step 6:** End

Fig. 3.26 Ant colony algorithm

3.3.1.14   Fuzzy Logic Control Algorithms

The advantage of these techniques is that they can operate with little precision input values and do not require a mathematical model with great precision [17–22, 58]. In accordance with Table 3.2, if the power (P_{pv}) increased, the operating point should be increased as well. However, if the power (P_{pv}) decreased, the voltage ($V_{pv,ref}$) should do the same. In this regulator, the range of each input variable and the output variable is divided into seven classes (Table 3.2).

The system based on MPPT fuzzy logic controller is composed of two inputs that are the error (E) and change in error (CE) [21] as shown in Fig. 3.28.

The bloc Simulink is in Fig. 3.29.

Some simulation results are given in Fig. 3.30.

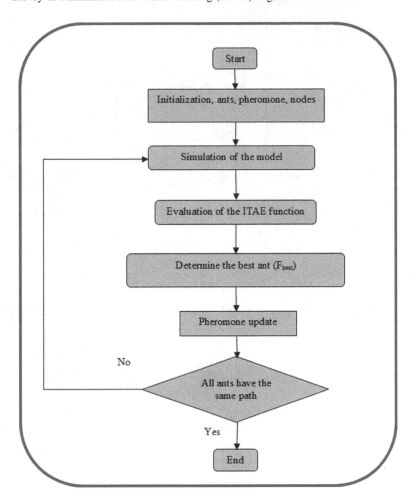

Fig. 3.27 Flowchart of the ACO algorithm

Table 3.2 Fuzzy rules table

Error (E)	Error variation (CE)						
	NG	NM	NP	ZE	PP	PM	PG
NG	NG	NG	NG	NG	NM	NP	ZE
NM	NG	NG	NG	NM	NP	ZE	PP
NP	NG	NG	NM	NP	ZE	PP	PM
ZE	NG	NM	NP	ZE	PP	PM	PG
PP	NM	NP	ZE	PP	PM	PG	PG
PM	NP	ZE	PP	PM	PG	PG	PG
PG	ZE	PP	PM	PG	PG	PG	PG

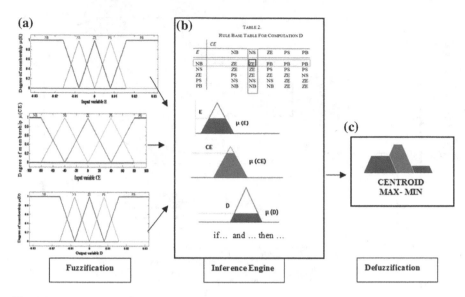

Fig. 3.28 Membership functions for: **a** input variable E, **b** input variable CE, **c** output variable D

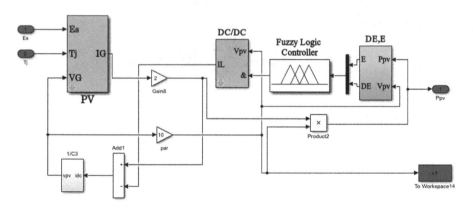

Fig. 3.29 FLC MPPT under MATLAB/Simulink

3.3.1.15   Adaptive Fuzzy Logic Controller (AFLC)

The AFLC is improved from scaling fuzzy logic controller (FLC), and it is mainly to adjust the duty cycle of the defuzzification of FLC for facing many kinds of external. Voltage VPV and current IPV of PV module are combined with the previous VPV and IPV for the averaged value.

The error (E) and the variation error (CE) of the system and of the modifier-based learning are used to modify the fuzzy parameters to optimize system operation. The controller Mamdani type with seven classes' membership functions is represented in Table 3.3. The errors are given by [10, 35]:

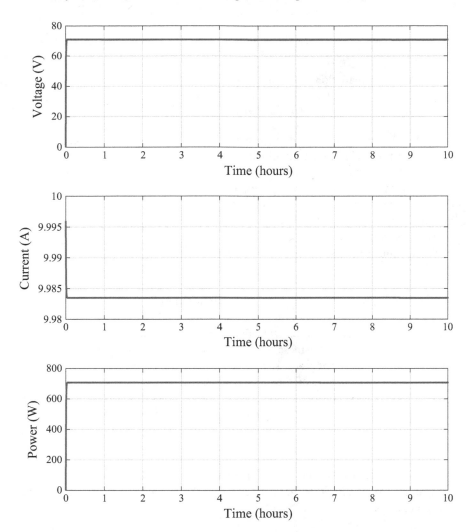

Fig. 3.30 Simulation results with FLC

Table 3.3 Modified fuzzy rules table

Error (E)	Variation error (CE)						
	NB	NM	NS	ZE	PS	PM	PM
NB	NB	NB	NM	ZE	ZE	ZE	ZE
NM	NB	NM	NM	ZE	NM	PS	PS
NS	NB	NB	NB	NB	PM	PS	PM
ZE	NB	NB	NS	ZE	PS	PM	PB
PS	NM	NS	ZE	PS	PM	PB	PB
PM	NS	PB	PB	PB	PB	PB	PB
PB	ZE	PB	PB	PB	PB	PB	PB

$$E(k) = \frac{P_{\text{pv}}(k+1) - P_{\text{pv}}(k)}{V_{\text{pv}}(k+1) - V_{\text{pv}}(k)} \tag{3.14}$$

And the error variation $CE(k)$ is

$$CE(k) = E(k+1) - E(k) \tag{3.15}$$

The fuzzy parameters can be adjusted using the following condition:
Si $E\langle\varepsilon$ (limit value), then the modifier-based learning will be selected.

The AFLC method is composed of two parts: The fuzzy logic control and adaptive mechanism. The FLC is one part of AFLC, which is composed of three units: fuzzification, fuzzy rules and defuzzification [3]. Figure 3.31 shows the membership function of AFLC method.

The bloc Simulink is in Fig. 3.32.

Some simulation results are represented in Fig. 3.33.

3.3.1.16   Sliding Mode Control (SMC) Algorithm

This method has various important advantages as high precision, good stability, simplicity, invariance, robustness, etc. The design of the control can be obtained in three important steps and each step is dependent on another one (the choice of surface, the establishment of the invariance conditions and determination of the control law). The structure of a controller by sliding mode consists of two parts; one concerns the exact linearization (u_{eq}) and the other stability u_n [48, 49].

$$u = u_{\text{eq}} + u_n \tag{3.16}$$

where u_{eq} corresponds to the control. It serves to maintain the variable control on the sliding surface and u_n is the discrete control determined to check the convergence condition.

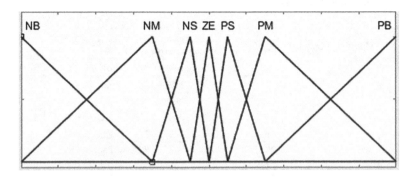

Fig. 3.31 Membership functions of AFLC method

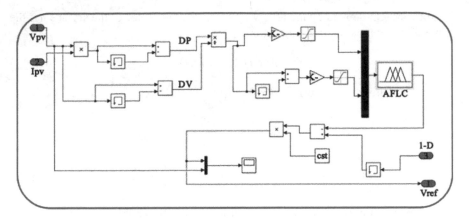

Fig. 3.32 AFLC model under MATLAB/Simulink

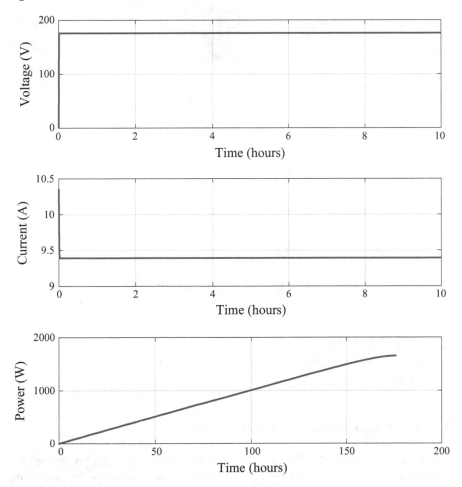

Fig. 3.33 Simulation results with AFLC

For a system defined by the following equation, the vector of the surface has the same dimension as the control vector (u).

$$\dot{x} = A(x,t) \cdot x + B(x,t) \cdot u \tag{3.17}$$

The derivative of the surface $S(x)$ is [5]:

$$\dot{S}(x) = \frac{\partial S}{\partial t} = \frac{\partial S}{\partial x} \cdot \frac{\partial x}{\partial t} \tag{3.18}$$

$$\dot{S}(x) = \frac{\partial S}{\partial x} \cdot (A(x,t) + B(x,t) \cdot u_{eq}) + \frac{\partial S}{\partial x} \cdot B(x,t) \cdot u_n \tag{3.19}$$

The expression of the equivalent control is then deduced:

$$u_{eq} = -\left(\frac{\partial S}{\partial t} \cdot B(x,t)\right)^{-1} \cdot \frac{\partial S}{\partial t} \cdot A(x,t) \tag{3.20}$$

For the equivalent control can take a finite value, it must:

$$\frac{\partial S}{\partial x} \cdot B(x,t) \neq 0 \tag{3.21}$$

The new expression of the surface derivative will be:

$$\dot{S}(x,t) = \frac{\partial S}{\partial x} \cdot B(x,t) \cdot u_n \tag{3.22}$$

And the condition becomes:

$$S(x,t) \cdot \frac{\partial S}{\partial x} \cdot B(x,t) \cdot u_n \langle 0 \tag{3.23}$$

The simplest form that can take the discrete control is as follows:

$$u_n = k_s \cdot \text{sign}(S(x,t)) \tag{3.24}$$

where the sign of k_s must be different from that of $\frac{\partial S}{\partial x} \cdot B(x,t)$.

The block under MATLAB/Simulink is in Fig. 3.34.
Simulation results are given in Fig. 3.35.

3.3.1.17 Neural Network Algorithm (ANN)

Neural network algorithm offers a great alternative for solving complex problems. It establishes relationships between input and output variables by analyzing

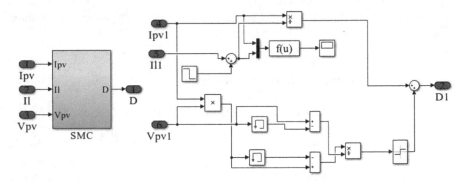

Fig. 3.34 Sliding mode control algorithm under MATLAB/Simulink

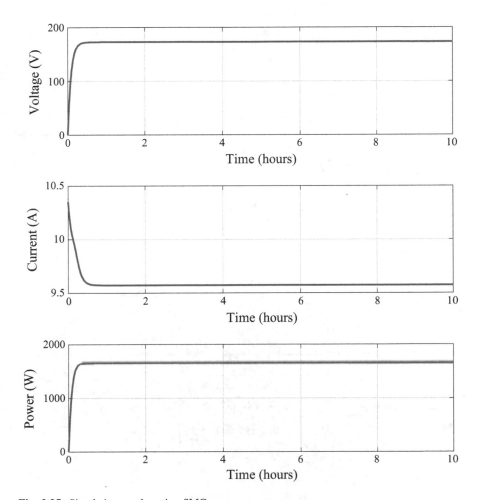

Fig. 3.35 Simulation results using SMC

previously stored data [15, 31]. Neural networks are generally composed of three different layers: Input layer, hidden layer and output layer (Fig. 3.36).

3.3.1.18   MPPT Using Neuro-fuzzy Network

The neuro-fuzzy controller has two inputs (e and Δe) and a single output (D), where e represents the error and Δe the error variation. The two input variables generate action control and adjust the duty cycle D to be applied to the DC/DC converter so as to ensure the adaptation of the power supplied by the GPV. This controller allows the automatic generation of fuzzy rules based on the Sugeno inference model [22]. The block diagram of a photovoltaic system with a MPPT control network-based neuro-fuzzy is represented in Fig. 3.37.

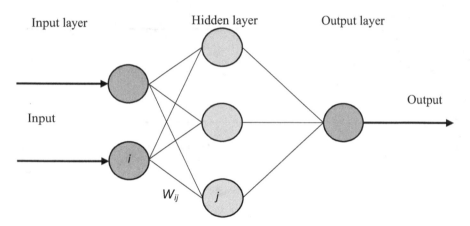

Fig. 3.36 Example of a neural network

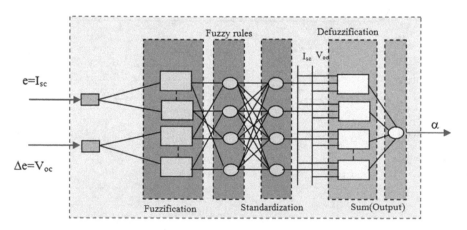

Fig. 3.37 Photovoltaic system with MPPT control by neuro-fuzzy network

The bloc Simulink is in Fig. 3.38.

The main advantage of this algorithm comparing to a conventional single ANN algorithm is the distinct generalization ability [67].

3.3.1.19 Genetic Algorithms

Genetic algorithms (GA) are stochastic optimization algorithms based on mechanisms of natural selection and genetics (Fig. 3.39). Its operation is extremely simple. It starts with an initial population which is encoded for the model of problem by some methods.

3.3.1.20 Hybrid Methods

Hybrid P&O and IncCond Methods (HP&OIncCond)

P&O method can be combined to IncCond method to improve performances (Fig. 3.40).

Simulation results under sudden solar radiation are given in Fig. 3.41.

Improved MPPT using FLC

A reference voltage (V_{ref}) based on solar irradiance and temperature for MPP is used as the input of the FLC. It is calculated in an adapted calculation (AC) block. The advantage of this method is that the number of membership functions in FLC decrease, so the tacking capability increase and the oscillations are reduced. The block diagram of this method is in Fig. 3.42.

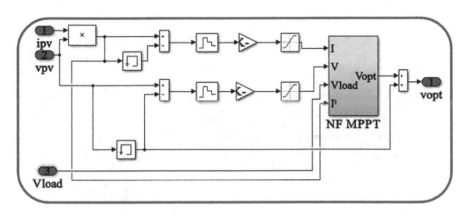

Fig. 3.38 Block diagram of NF MPPT

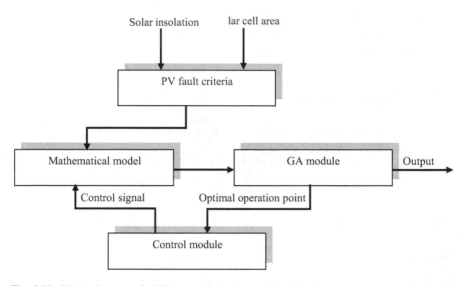

Fig. 3.39 Block diagram of a PV system based on genetic algorithm

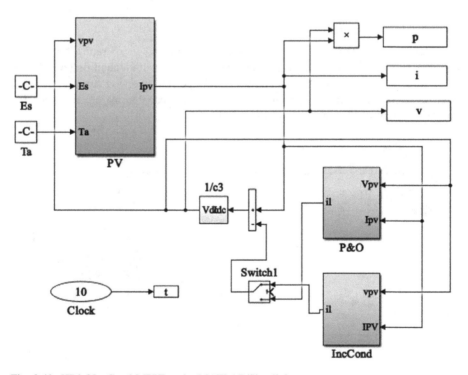

Fig. 3.40 HP&OIncCond MPPT under MATLAB/Simulink

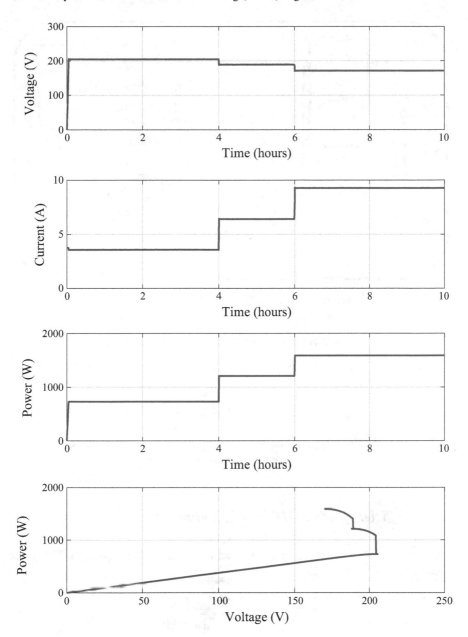

Fig. 3.41 Simulation results with hybrid P&O/IncCond method

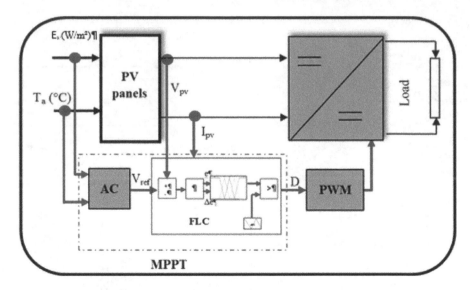

Fig. 3.42 Block diagram of improved MPPT using FLC

Indirect Hybrid Fuzzy-P&O Variable Step Size MPPT Controller

Authors in [68] have proposed variable step size for hybrid fuzzy-P&O MPPT algorithm. It is given as:

$$D(k) = D(k-1) + F.S + A.D \qquad (3.25)$$

With $D(k)$ is the duty cycle at the instant k, $D(k-1)$ the duty cycle at the instant $(k+1)$, $A.D$ the fixed step size and $F.S$ is the scaling factor.

3.3.2 Efficiency of an MPPT Algorithm

Efficiency η_{MPPT} is the most important parameter of an MPPT algorithm. This value is calculated as:

$$\eta_{\mathrm{MPPT}} = \frac{\int_0^t P_{\mathrm{pv-max}}(t)\mathrm{d}t}{\int_0^t P_{\mathrm{pv-MPPT}}(t)\mathrm{d}t} \qquad (3.26)$$

where $P_{\mathrm{pv-MPPT}}$ represents the output power of PV system with MPPT, and $P_{\mathrm{pv-max}}$ is the output power at true maximum power point.

3.3.3   Comparison of Different Algorithms

The P&O is the simplest MPPT method but its disadvantage is that there are oscillations around the MPP in the steady state. To overcome this problem different techniques have been proposed in the literature. The most used are related in Table 3.4 [47, 48, 54, 59].

The disadvantage of IncCond method with fixed step is its oscillations, so improved methods have appeared [47, 48, 54, 59] (Table 3.5).

Other simple MPPT methods are cited in paper conferences and publications [47, 48, 54, 59]. Generally, they all depend on PV array, used only one sensor and no accurate (Table 3.6).

Table 3.4 Classical and advanced P&O algorithms

MPPT method	Symbol	Properties
Perturb and observe	P&O	– Most used, a well-known iterative method and simple structure – The number of input variables (I_{pv}, V_{pv}) – Cost: low – Implementation: simple – No dependency of PV array – Good tracking factor – Accurate – Two sensors (current and voltage) – Analogic/digital circuits – For applications that do not require high accuracy – Oscillations around the MPP in the steady state – Sometimes leads to a wrong tracking direction to the MPP
Improved dP-P&O algorithm	dP-P&O	– Good tracking speed – Track in the right direction under rapid variation of solar irradiance – No dependency of PV array – Analogic/digital circuits – Two sensors (current and voltage)
Modified perturb and observe	MP&O	– Implementation: complex – No dependency of PV array – Very good tracking factor – Accurate – Two sensors (current and voltage)
Optimized P&O algorithm	OP&O	– Variable step size is used
Estimate, perturb and observe method	EP&O	– Good tracking speed – Same tracking accuracy as P&O method
Combined P&O with CV	P&OCV	– Track the MPP at low and high solar irradiance – High efficiency – Complex

Table 3.5 Classical and modified algorithms IncCond

MPPT methods	Symbol	Proprieties
Incremental conductance	Inc/ IncCond	– The number of input variables (I_{pv}, V_{pv}) – Cost: medium – Implementation: medium – No dependency of PV array – Good tracking factor – Accurate – Two sensors (current and voltage) – Analogic/digital circuits – For applications that do not require high accuracy
Modified Incremental conductance	MInc	– No dependency of PV array – Very good tracking factor – Implementation: complex – Accurate – Two sensors (current and voltage)

Another MPPT method category needs two sensors, so more accurate and with a medium complexity are cited [47, 48, 54, 59] (Table 3.7).

Finally, intelligent methods have improved classical ones with more accuracy but still complex in implementation, so the cost is high (Table 3.8).

A comparison with some MPPT methods is given in Fig. 3.43.

The simulation results obtained by the different MPPT methods are almost the same. Advanced controllers (fuzzy and sliding mode controllers) are faster and more robust in the case of slow variation of solar radiation, with an advantage for the adaptive fuzzy controller.

3.3.4 Global Maximum Power Point Tracking (GMPPT) Techniques of Photovoltaic System Under Partial Shading Conditions

Shading one panel decreases the power of the entire system. All shadows have to been considered, including those produced by small obstacles. Partial shading (PSH) can be caused by several factors such as dust, sand, tree leaves, bird litters and new constructions, especially after a photovoltaic installation. Different works in literature have been focused on this problem of PSH and different solutions have been proposed [36–39]. Most of these articles solve the partially shaded situation by introducing maximum power point tracking (MPPT) techniques. These solutions sometimes require more advanced measurements for photovoltaic arrays, so higher costs and a longer time to find the maximum power point (MPP). But they allow increasing the PV power. Over the years, several MPPT algorithms have been developed and widely adapted to determine the MPP [57].

Table 3.6 Other MPPT methods based on only or without sensor

MPPT method	Symbol	Proprieties
Fixed duty cycle	FDC	– No dependency of PV array – Poor tracking factor – Implementation: simple – No accurate – No sensors
Constant reference voltage/constant voltage	CRV/ CV	– Dependency of PV array – Reasonable tracking factor – Implementation: simple – No accurate – Only a voltage sensor is used – Economical – Cost: low
Fractional short-circuit current	FSCC	– Dependency of PV array – No accurate – Efficiency: low – Both analogic and digital circuits used – Simplicity – Implementation: low cost – Only a current sensor is used
Fractional short-circuit voltage	FSCV	– Dependency of PV array – No accurate – Easy method – Low cost – Fast operation is achieved – Both analogic and digital circuits used – Simplicity – Only a voltage sensor is used
One-cycle control technique	OCC	– Dependency of PV array – Implementation: medium – Both analogic and digital circuits used – Only a current sensor is used

When there is partial shading, the system has multi-peak power characteristics. In order to properly track the global maximum power point (GMPP), an effective method is necessary. Different types of MPPT techniques have been used to detect global MPP. The most important are the PSO, artificial bee colony, ant colony optimization, hybrid MPPT as P&O with neural network, particle swarm optimization with P&O, etc.

Table 3.7 MPPT based on two sensors

MPPT method	Symbol	Proprieties
Colony of firefly algorithm	CFA	– Accurate – Inexpensive – Complexity: medium
Bayesian network	BN	
Array reconfiguration	AR	– Simple – Accurate – Dependency of PV array – Expensive – Implementation medium – Digital circuits – Two sensors (current and voltage)
Steepest descent method	STD	– Accurate – Complexity: medium – Expensive – Two sensors (current and voltage)
Load current/Load voltage maximization	LCM/ LVM	– No dependency of PV array – Two sensors (current and voltage) – Analogic circuits used – Implementation: medium
Switching ripple correlation control	SRCC	– No dependency of PV array – Good tracking factor – Implementation: complex – Accurate – Two sensors (current and voltage) – The time taken to achieve MPP is less – Analogic circuits used
Current sweep method	CW	– A sweep waveform for the array is used – Complex implementation – Dependency of PV array – Two sensors (current and voltage)
DC-link capacitor droop control technique	DM	– Simple – No dependency of PV array – Implementation: low – Voltage sensor – Both analogic and digital circuits used

3.3.4.1 PV Model with Shading

The shading fault can be modeled by the different parameters variation of the cell. When the components are connected in series, the voltage produced by each component is no longer equal for the same current. And when the components are connected in parallel, the current supplied by each component is no longer the same for the same voltage [57].

Table 3.8 Advanced MPPT algorithms

MPPT method	Symbol	Proprieties
Particle swarm optimization	PSO	– Simple structure – Fast computation ability – Easy implementation
Ant colony optimization	ACO	– Complex
Fuzzy logic control	FLC	– The number of input variables (I_{pv}, V_{pv}) – Cost: high – Implementation: high difficulty – No dependency of PV array – Good tracking factor – Accurate – Two sensors (current and voltage) – Analogic circuits – For applications requiring high accuracy – Dependency of PV array
Adaptive fuzzy logic control	AFLC	– The number of input variables (I_{pv}, V_{pv}) – Cost: high – Implementation: high difficulty – $\xi_{MPPT} = 99.85\%$ – No dependency of PV array – Good tracking factor – Accurate – Two sensors (current and voltage) – Analogic circuits – For applications requiring high accuracy
Sliding mode control	SMC	– The number of input variables (I_{pv}, V_{pv}) – Cost: high – Implementation: high difficulty – No dependency of PV array – Good tracking factor – Accurate – Two sensors (current and voltage) – Analogic circuits – For applications requiring high accuracy – Dependency of PV array
Beta technique	BT	– Dependency of PV array – Excellent tracking factor – Implementation: medium – Accurate – Two sensors (current and voltage)
Artificial neural networks	ANN	– The number of input variables (I_{pv}, V_{pv}) – Cost: high – Implementation: high difficulty – No dependency of PV array – Good tracking factor – Accurate – Two sensors (current and voltage) – Analogic circuits – For applications requiring high accuracy

(continued)

Table 3.8 (continued)

MPPT method	Symbol	Proprieties
		– Dependency of PV array
Quadratic maximization	QM	– Use the quadratic polynomial to perform a MPP tracker – The second-order Lagrange interpolating polynomial is used
Sinusoidal extremum-seeking control	ESC	– Better performance than P&O – Reduces oscillation around the MPP

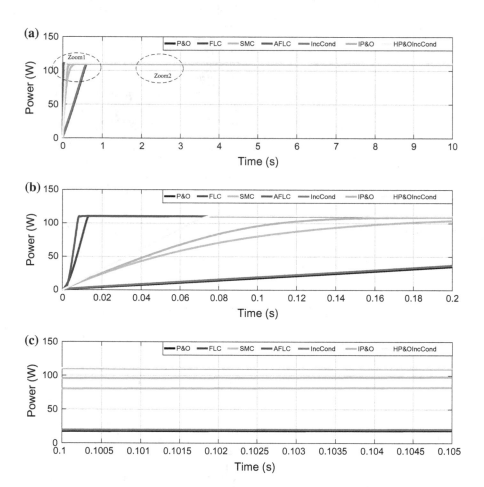

Fig. 3.43 Comparison of most important MPPT methods

The relationship of current and voltage of the ith cell of a group is given by:

$$I_{\text{cell}} = I_{\text{imposed}}$$
$$I_{\text{cell}} \xrightarrow{f(I_{\text{cell},i}, V_{\text{cell},i})=0} V_{\text{cell},i}$$

(3.27)

In the case of mismatch, for a given current, the voltage produced by the cells is not necessarily identical because their parameters are not the same. In the case of a group of cells, the sum of the voltages of the cells in the group can be negative. This is due to the fact that one or more cells in the group produce a negative voltage when they are crossed by a current higher than their short-circuit current. It is in this situation that the bypass diode plays its role when the total sum of the cell voltage becomes negative and thus derives the excess current for the shaded cell.

Equation (3.28) gives the current and the voltage of a jth cell of a group.

$$I_{\text{Group},j} = I_{\text{cell}} + I_{\text{bypass}}$$
$$V_{\text{Group},j} = \sum_{i=1}^{N_{\text{cell},i}} V_{\text{cell},i} \xrightarrow{\text{if}} \sum_{i=1}^{N_{\text{cell},i}} V_{\text{cell},i} \geq 0$$
$$V_{\text{Group},j} = 0 \xrightarrow{\text{if}} \sum_{i=1}^{N_{\text{cell},i}} V_{\text{cell},i} \langle 0$$

(3.28)

Equation (3.29) gives the relationship of the current and the voltage of kth module of a string.

$$I_{\text{Module},k} = I_{\text{Group}}$$
$$V_{\text{Module},k} = \sum_{j=1}^{N_{\text{Group}}} V_{\text{Group},j}$$

(3.29)

Equation (3.30) gives the relationship of the current and the voltage of the zth string of the fields.

$$I_{\text{String},Z} = I_{\text{Module}}$$
$$V_{\text{String},Z} = \sum_{K=1}^{N_{\text{Module}}} V_{\text{Module},k}$$

(3.30)

During the health conditions of solar irradiation, there is one peak in $P_{\text{pv}}(V_{\text{pv}})$ characteristic. But during partial shading, it can have multiple peaks (Fig. 3.44). This is due to the bypass diode. Global peak is defined as the maximum power of PV string and local peak appeared during partial shading.

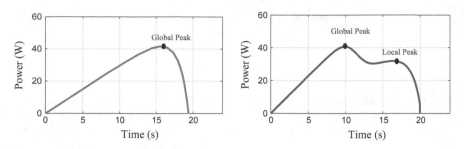

Fig. 3.44 $P_{pv}(V_{pv})$ characteristic with partial shading

3.3.4.2 Example Under Matlab/Simulink

An example has been made under MATLAB/Simulink (Fig. 3.45).
Electrical characteristics during partial shading are in Fig. 3.46.

3.4 MPPT Algorithms in Wind Turbine Systems

The output power of wind energy system varies depending on the wind speed. Due to the nonlinear characteristic of the wind turbine, it is difficult to maintain the maximum power output of the wind turbine for all wind speed conditions. Therefore, over the years, several maximum power point tracking (MPPT) algorithms have been developed to track the maximum power point of the wind turbine [2].

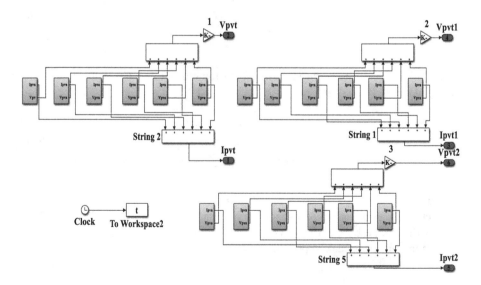

Fig. 3.45 Partial shading in MATLAB/Simulink

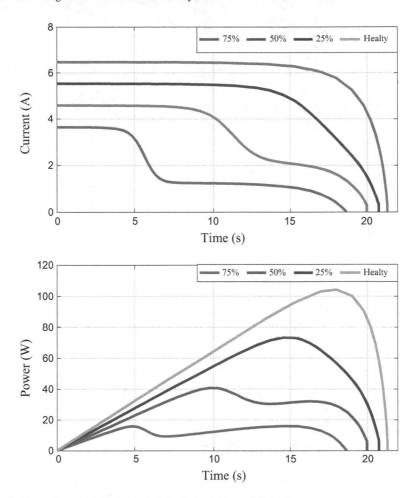

Fig. 3.46 $I_{pv}(V_{pv})°$ and $P_{pv}(V_{pv})$ characteristics with partial shading

3.4.1 Tip Speed Ratio Method (TSR)

In order to have the maximum possible power, the turbine should always operate at λ_{opt}. The TSR control method regulates the tip speed ratio to maintain it to an optimal value, at which the rotational speed is optimum and the power extracted is maximal. This control requires the knowledge of wind speed, the turbine speed and the reference optimal point of the TSR which can be determined experimentally or theoretically. The comparison of the TSR reference with the actual value feeds this difference to the controller and gives the reference power.

The expression of power as a function of turbine speed is:

$$P_t = \frac{1}{2}\rho C_p(\lambda)S\frac{R^3}{\lambda^3}\omega_t^3 \tag{3.31}$$

The electromagnetic torque is as follow:

$$T_t = \frac{1}{2}C_p\rho\pi\frac{R^5}{\lambda^3}\omega_t^2 \tag{3.32}$$

Considering the optimal conditions the torque will have the following form:

$$T_{\text{emopt}} = K_{\text{opt}}\omega_t^2 \tag{3.33}$$

with

$$K_{\text{opt}} = \frac{1}{2}C_p\rho\pi\frac{R^5}{\lambda^3} \tag{3.34}$$

The MPPT algorithm, using the measured rotational speed, determines the reference torque shown in Fig. 3.47.

Figure 3.48 shows the block diagram of a WECS with TSR method.

We make an application of the TSR method in a wind system. The block diagram under MATLAB/Simulink can be represented in Fig. 3.49.

3.4.2 Power Signal Feedback (PSF) Method

Power signal feedback method generates a reference power signal to maximize the output power. However, it requires the knowledge of the wind turbine and maximum power curve which can be obtained from the experimental results or simulations [2]. Then, the data points for maximum turbine power and the corresponding wind turbine speed must be recorded in a look-up table. The PSF control method regulates the turbine power to maintain it to an optimal value, so that the power coefficient C_p is always at its maximum value corresponding to the optimum tip speed ratio. Figure 3.50 shows the block diagram of a WECS with TSR control.

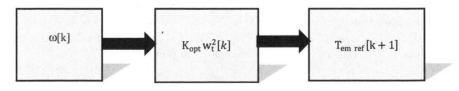

Fig. 3.47 Reference torque as a function of speed

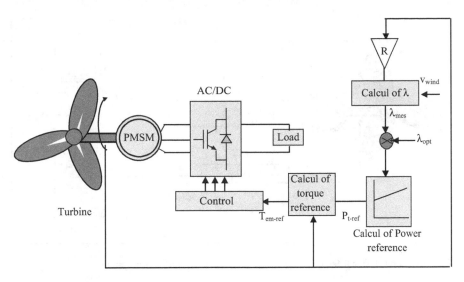

Fig. 3.48 Tip speed ratio method of wind energy conversion system

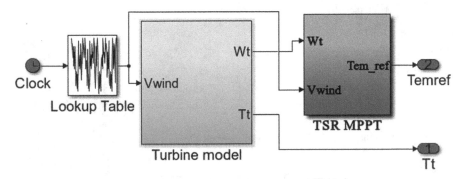

Fig. 3.49 Block diagram of WTCS with TSR MPPT method

The block diagram of WTCS with PSF method under MATLAB/Simulink is represented in Fig. 3.51.

The PSF algorithm is in Fig. 3.52.

3.4.3   Optimal Torque Control (OTC)

Optimal torque control is a slight variant of PSF control [26]. It adjusts the generator torque to its optimal at different wind speeds. However, it requires the knowledge of turbine characteristics (C_{pmax} and λ_{opt}).

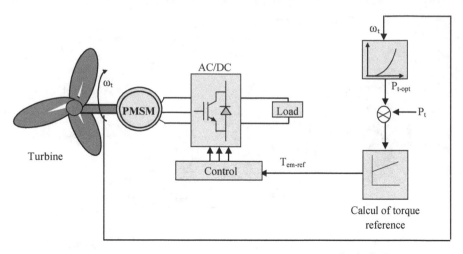

Fig. 3.50 Power signal feedback control of wind energy conversion system

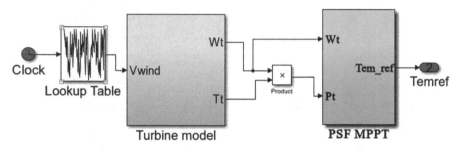

Fig. 3.51 Block diagram of WTCS with PSF method

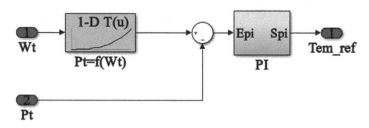

Fig. 3.52 PSF under MATLAB/Simulink

We have:

$$v_{\text{wind}} = \frac{R.\omega_t}{\lambda} \tag{3.35}$$

We obtain the power function of rotational turbine speed

$$P_t(\omega_t) = \frac{1}{2} \cdot \frac{C_p(\omega_t).\rho.\pi.R^4}{\lambda^3(\omega_t)} .\omega_t^3 \tag{3.36}$$

With:

$$P_t(\omega_t) = T_{\text{em}}.\omega_t \tag{3.37}$$

$$C_t.\omega_t = \frac{1}{2} .C_p(\lambda).\rho.\pi.R^2.v_v^3 \tag{3.38}$$

Then:

$$T_{\text{em}} = \frac{1}{2} \cdot \frac{C_p(\omega_t).\rho.\pi.R^4}{\lambda^3(\omega_t)} .\omega_t^2 \tag{3.39}$$

Assuming optimal conditions, power value, speed and optimal torque are given by the following relations:

$$P_{\text{opt}} = \frac{1}{2} .C_{\text{p-opt}}.\rho.\pi.R^2.v_{\text{wind}}^3 \tag{3.40}$$

$$\omega_{\text{opt}} = \frac{v_{\text{wind}}.\lambda_{\text{opt}}}{R} \tag{3.41}$$

$$T_{\text{em-opt}} = T_{\text{em-ref}} = K_{\text{opt}}.\omega_{\text{opt}}^2 \tag{3.42}$$

With:

$$K_{\text{opt}} = \frac{1}{2} \cdot \frac{C_{\text{p-opt}}(\omega_t).\rho.\pi.R^4}{\lambda_{\text{opt}}^3(\omega_t)} \tag{3.43}$$

Figure 3.53 shows the block diagram of a WECS with optimal torque control. It can be implemented in MATLAB/Simulink as in Fig. 3.54.

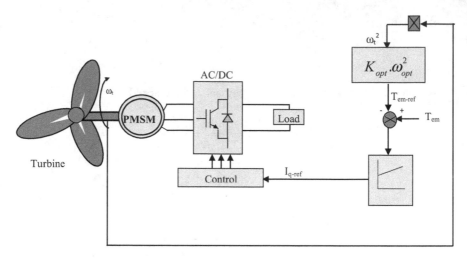

Fig. 3.53 Optimal torque control of wind energy conversion system

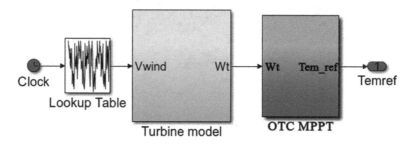

Fig. 3.54 OTC MPPT under MATLAB/Simulink

3.4.4 Hill Climb Searching (HCS) Technique

HCS technique is usually used for the peak power of the wind turbine that will maximize the extracted energy [12]. This control efforts to climb the $P_t(\omega_t)$ curve in the direction of increasing P_t, by varying the rotational speed periodically with a small incremental step to reduce the oscillation around the MPP, the P&O algorithm compares the power previously delivered with the one after disturbance (Fig. 3.55).

It is implemented in MATLAB/Simulink as in Fig. 3.56.

3.4.5 MPPT Based on Gradient Method (GM)

The gradient method is very simple. It is not necessary to know the parameters λ_{opt} and C_{pmax} for each wind speed. The speed reference is adjusted in order to make the

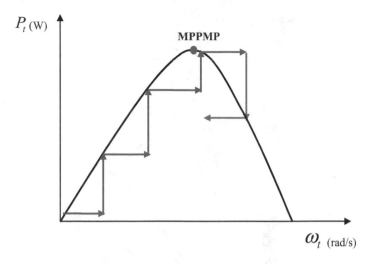

Fig. 3.55 HCS principle

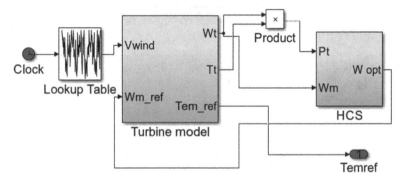

Fig. 3.56 HCS MPPT under MATLAB/Simulink

turbine operates around the point that provides the maximum power for each wind speed. To adjust the rotational speed of the generator, the variation direction of the ratio $\frac{dP_t}{d\omega_t}$ is controlled. When this ratio will be zero, the desired maximum power is reached [47].

$$\frac{dP_t}{d\omega_t} = \frac{dP_t}{dt} \cdot \left(\frac{d\omega_t}{dt}\right)^{-1} \tag{3.44}$$

Four possible cases can be possible (Table 3.9) and illustrated in Fig. 3.57.

The algorithm requires knowledge of power and speed at all times. According to the four possible cases described in Table 3.9, at the iteration k, the speed reference

Table 3.9 Possible cases for GM

$\dfrac{d\omega_t}{dt}$	$\dfrac{dP_t}{dt}$	
	$\rangle 0$	$\langle 0$
$\rangle 0$	Case 1: $\omega_t \uparrow$	Case 2: $\omega_t \downarrow$
$\langle 0$	Case 3: $\omega_t \downarrow$	Case 4: $\omega_t \uparrow$

Fig. 3.57 Gradient method principle

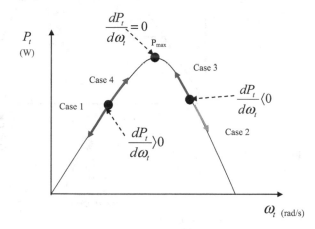

$\Delta\omega_t\,(k+1)$ is increased or decreased by a fixed step $\Delta\omega_t$. The optimization algorithm is illustrated by the flowchart in Fig. 3.58.

It can be implemented under MATLAB/Simulink as in Fig. 3.59.

3.4.6 Hybrid MPPT Technique HP&O/OTC

P&O and OTC MPPT methods present disadvantages but the combination of the two algorithms allows us to obtain a hybrid MPPT HP&O/OTC (Fig. 3.60).

The application under MATLAB/Simulink can be represented in Fig. 3.61.

3.4.7 Fuzzy Logic Controller Technique

Rules to be observed in order to converge toward the optimal point are relatively simple to establish. These rules depend on the wind power variations ΔP_t and speed variation $\Delta\omega_t$, which give a torque reference Te_{ref} and a wind turbine reference speed ω_{ref} (Fig. 3.62).

The fuzzy rules used for the fuzzy controller are given in Table 3.10.

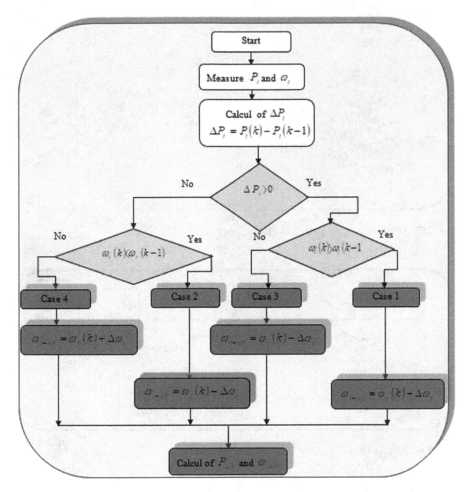

Fig. 3.58 Flowchart of GM

3.4.8 *Artificial Neural Networks (ANN) Method*

Artificial neural networks (ANN) are electronic models based on the neural structure of the brain. This function permits ANNs to be used in the design of adaptive and intelligent systems since they are able to solve problems from previous examples. ANN models involve the creation of massively paralleled networks composed mostly of nonlinear elements known as neurons. Each model involves the training of the paralleled networks to solve specific problem [15]. ANNs consist of neurons in layers, where the activations of the input layer are set by an external parameter. Generally, networks contain three layers—input, hidden and output (Fig. 3.63). The input layer receives data usually from an external source while the output layer sends information to an external device. There may be one or more

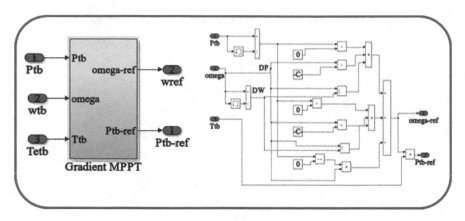

Fig. 3.59 Gradient method under MATLAB/Simulink

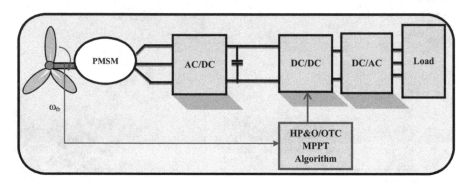

Fig. 3.60 Hybrid HCS/OTC MPPT in wind turbine system

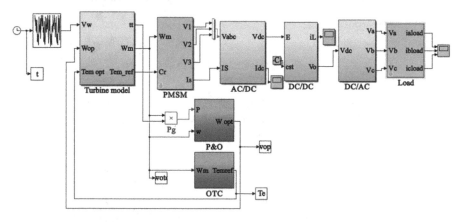

Fig. 3.61 Application of HP&O/OTC in stand-alone wind turbine system

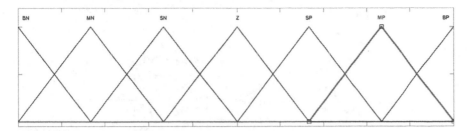

Fig. 3.62 Membership of the error (E), the error change (CE) and the output ω_{ref}

Table 3.10 Fuzzy rules used for the fuzzy controller

Error (E)	Error variation (CE)						
	NB	NM	NS	ZE	PS	PM	PM
NB	NB	NB	NB	NB	NM	NS	ZE
NM	NB	NB	NB	NM	NS	ZE	PS
NS	NB	NB	NM	NS	ZE	PS	PM
ZE	NB	NM	NS	ZE	PS	PM	PB
PS	NM	NS	ZE	PS	PM	PB	PB
PM	NS	ZE	PS	PM	PB	PB	PB
PB	ZE	PS	PM	PB	PB	PB	PB

hidden layers between the input and output layers. The back-propagation method is the common type of learning algorithm [48].

3.4.9 Radial Basis Function Network (RBFN)

Radial basis function network has a similar feature to fuzzy system. The output value is calculated using the weighted sum method and the number of nodes in the hidden layer of the RBFN is the same as the number of if-then rules in the fuzzy system. The receptive field functions of the RBFN are similar to the membership functions of the premise part in the fuzzy system. An application of RBFN on a wind energy conversion system is represented in Fig. 3.64. The electrical generator is driven by a wind turbine supplying the power to a load (grid for example), through back to back converters. To control the DC/AC converter, we use an MPPT (P&O and RBFN) control for maximizing power and a PWM control. The reference dc voltage V_{dcref} is obtained using P&O method and RBFN controller force V_{dc} to follow its reference V_{dcref} and adjust the load current reference I_{Loadref} for the PWM inverter control.

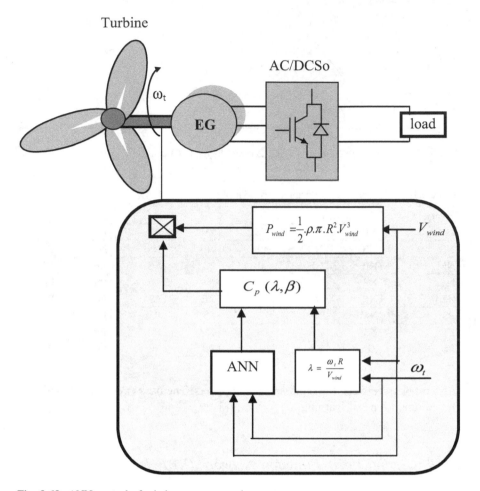

Fig. 3.63 ANN control of wind energy conversion system

3.4.10 Adaptative Neuro-fuzzy Inference System (ANFIS)

The ANFIS controller is designed and adapted to tracking a maximum power of the wind. ANFIS is the integration of artificial neural networks and fuzzy inference systems [2]. Neural network (NN) is used to adjust input and output parameters of membership function in the fuzzy logic controller (FLC). Atypical architecture of a neuro-fuzzy network for two inputs (*x* and *y*) is shown in Fig. 3.65.

The first layer is called input layer. Each node of this layer stores the parameters to define a bell-shaped membership function. In the second layer, each node performs connective operation "AND" within the rule antecedent to determine the

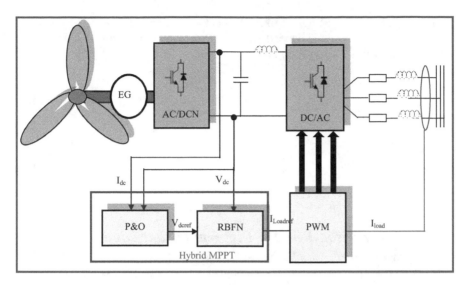

Fig. 3.64 Wind energy conversion system with RBFN controller

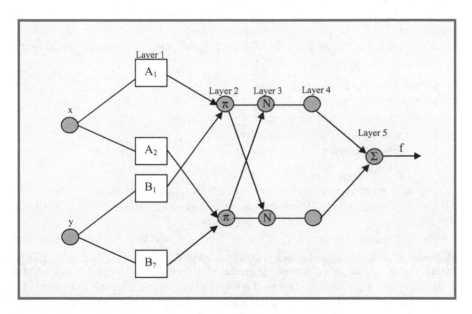

Fig. 3.65 General structure of neuro-fuzzy controller

corresponding firing, the nodes of layer three perform a normalization process to produce the normalized firing strength. The fourth layer deals with the consequent part of the fuzzy rule. The node of this layer is adaptive with output. Finally, the fifth layer which is the final output is the weighted average of all rule outputs [48].

The first of three input signals of ANFIS is the error signal $e(t)$, the second one is the changing of error signal depending on time $de(t)/dt$ and the third input signal of ANFIS is the value of power P_{mec}.

We have:

$$e(t) = \omega_t - \omega_{t-\text{nom}} \tag{3.45}$$

And:

$$\frac{de(t)}{dt} = \frac{\omega_t - \omega_{t-\text{nom}}}{dt} \tag{3.46}$$

3.4.11 Comparison Between Different Optimization Methods

A comparison of some MPPT algorithms, in terms of electromagnetic torque and rotational speed, has been made and the different results are (Fig. 3.66).

The HCS or P&O is the simplest method which not requires the knowledge of parameters except the power. To make sure that conditions changes do not lead HCS method in the wrong direction, different techniques have been proposed in the literature. The most used are [54, 59]:

- HCS with variable step size;
- HCS with dual step size;
- HCS with search-remember-reuse;
- HCS modified to avoid generator stall;
- HCS with disturbance injection band;
- HCS with limit cycle, etc.

The TSR method provides the fastest control action. An improved method of TSR known as adaptive TSR control has been applied in wind turbine. In this case, the TSR value is not constant but turned online.

The PSF method is based on look-up table. A modified PSF method is also proposed to avoid generator stall. The OTC algorithm uses the quadratic optimal torque curve to maximize power. There is no difference between PSF and OTC methods in the performances and in the implementation. There is a modified OTC method used for fast training. Finally, methods based on artificial intelligence are more effective and give better performances but require speed sensors, one for the wind speed and another for the generator speed [26].

An overview of some important proprieties of the most used MPPT methods has been summarized in Table 3.11 [47, 48, 54, 59].

Fig. 3.66 Comparison of
most used MPPT in wind
turbine system

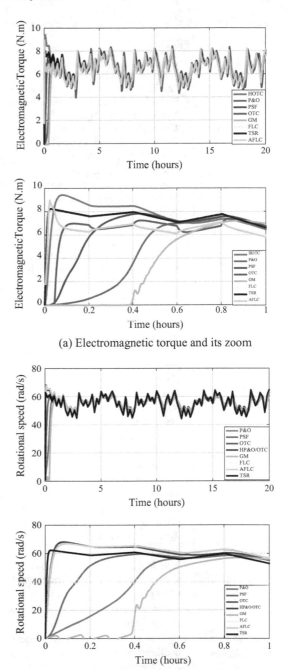

Fig. 3.66 Comparison of most used MPPT in wind turbine system

(a) Electromagnetic torque and its zoom

(b) Rotational speed and its zoom

Table 3.11 Some important proprieties of the most used MPPT methods

MPPT methods	Proprieties
TSR	– Regulates the wind turbine rotor speed to maintain an optimal tip speed ratio – The TSR direction control method is limited by the difficulty in wind – Speed and turbine speed measurements – Needs wind and rotor speed – The implementation is high – Require pre-known value of optimal tip speed ration – More costly (02 sensors)
PSF	– It is based on a look-up table – Requires the knowledge of the wind turbine's maximum power curve and tracks this curve through its control mechanisms – Needs rotor speed – The implementation is medium
OTC	– Simple and accurate – Use the quadratic optimal torque curve to maximize power – Adjusts the generator torque to its optimal at different wind speeds – Need only measured rotor speed
HCS or P&O	– Its purpose is to search continuously for the peak output power of the wind turbine – It is a popular method due to its simplicity and independence of system characteristics – It can be fast and effective in spite of the variations in wind speeds and the presence of turbine inertia – No require knowing parameters – It is simple to implement
SMC	– High precision – Good stability – Simplicity, invariance – Robustness, etc.
FLC	– Can reduce the effect of cutting force disturbances – No require knowing parameters – Require speed sensors
AFLC	– Can perform an adaptive fuzzy inference process using various inference parameters, such as the shape and location of a membership function, dynamically and quickly
PSO	– It applies an analogy of swarm behavior of natural creatures (birds or fish) – It finds the optimal solution using a population of particles and forces the system to reach its equilibrium quickly where the turbine inertia effect is minimized
WRBFM	– Maintain the system stability and reach the desired performance even with parameter uncertainties
ANN	– Do not require mathematical models and have the ability to approximate nonlinear systems – Effective and robust – Require a long offline training – Needs 02 sensors

(continued)

Table 3.11 (continued)

MPPT methods	Proprieties
RBFM	– It has a similar feature to the fuzzy system – It is very useful to be applied to control the dynamic systems
ANFIS	– It is the integration of artificial neural networks and fuzzy inference systems
GM	– Simple method – Requires knowledge of power and speed at all times

3.5 Conclusion

This chapter has been devoted to the optimization power of PV, wind and hybrid systems. The most used MPPT algorithms by students have been presented with some details in MATLAB/Simulink.

References

1. Amer M, Namaane A, M'Sirdi NK (2013) Optimization of hybrid renewable energy systems (HRES) using PSO for cost reduction. Energy Procedia 42:318–327. https://doi.org/10.1016/j.egypro.2013.11.032
2. Rekioua D (2014) Optimisation of wind system conversion. Green energy and technology, pp 77–105. 9781447164241
3. Rekioua D (2014) Wind energy conversion and power electronics modeling. Green energy and technology, pp 51–57. 9781447164241
4. Rahrah K, Rekioua D, Rekioua T (2015) Optimization of a photovoltaic pumping system in Bejaia (Algeria) climate. J Electr Eng 15(2):321–326
5. Achour A, Rekioua D, Mohammedi A, Mokrani Z, Rekioua T, Bacha S (2016) Application of direct torque control to a photovoltaic pumping system with sliding-mode control optimization. Electr Power Compon Syst 44(2):172–184
6. Mokhtari Y, Rekioua D (2018) High performance of maximum power point tracking using ant colony algorithm in wind turbine. Renew Energy 126:1055–1063
7. Yang XS (2008) Nature-inspired metaheuristic algorithms. Luniver Press
8. Fathabadi H (2016) Novel highly accurate universal maximum power point tracker for maximum power extraction from hybrid fuel cell/photovoltaic/wind power generation systems. Energy 116:402–416
9. Kumar MBH, Saravanan B, Sanjeevikumar P, Blaabjerg F (2018) Review on control techniques and methodologies for maximum power extraction from wind energy systems. IET Renew Power Gener 12(14):1609–1622, art. no. Y
10. Yaakoubi AE, Amhaimar L, Attari K, Harrak MH, Halaoui ME, Asselman A (2019) Non-linear and intelligent maximum power point tracking strategies for small size wind turbines: Performance analysis and comparison. Energy Rep 5:545–554
11. Karabacak M, Fernandez-Ramirez LM, Kamal T, Kamal S (2019) A new hill climbing maximum power tracking control for wind turbines with inertial effect compensation. IEEE Trans Ind Electron 66(11):8545–8556, 8681269

12. Mousa HHH, Youssef A-R, Mohamed EEM (2019) Adaptive P&O MPPT algorithm based wind generation system using realistic wind fluctuations. Int J Electr Power Energy Syst 112:294–308
13. Karabacak M (2019) A new perturb and observe based higher order sliding mode MPPT control of wind turbines eliminating the rotor inertial effect. Renew Energy 807–827
14. Kumar D, Chatterjee K (2017) Design and analysis of artificial bee-colony-based MPPT algorithm for DFIG based wind energy conversion systems. Int J Green Energy 14(4): 416–429
15. Tiwari R, Kumar K, Neelakandan RB, Padmanaban S, Wheeler PW (2018) Neural network based maximum power point tracking control with quadratic boost converter for PMSG— wind energy conversion system. Electronics (Switzerland) 7(2), art. no. 20
16. Rahrah K, Rekioua D, Rekioua T, Bacha S (2015) Photovoltaic pumping system in Bejaia climate with battery storage. Int J Hydrogen Energy 40(39):13665–13675
17. Kandemir E, Borekci S, Cetin NS (2018) Comparative analysis of reduced-rule compressed fuzzy logic control and incremental conductance MPPT methods. J Electron Mater 47 (8):4463–4474
18. Abdourraziq MA, Abdourraziq S, Maaroufi M (2018) Efficiency optimization of a microcontroller-based PV system employing a single sensor fuzzy logic controller. IET Power Electron 11(3)
19. Samal S, Barik PK, Sahu SK (2018) Extraction of maximum power from a solar PV system using fuzzy controller based MPPT technique. In: International conference on technologies for smart city energy security and power: smart solutions for smart cities, ICSESP 2018— proceedings, Jan 2018, pp 1–6
20. Alex LT, Dash SS, Sridhar R (2018) Fuzzy controller-aided input series output series (ISOS) modular converters for photovoltaic applications. Int J Fuzzy Syst 20(8):2566–2576
21. Bensmail S, Rekioua D, Serir C (2019) Optimization of a photovoltaic pumping system by applying fuzzy control type-1 with adaptive gain. Lect Notes Netw Syst 62:321–328
22. Chekired F, Larbes C, Rekioua D, Haddad F (2011) Implementation of a MPPT fuzzy controller for photovoltaic systems on FPGA circuit. Energy Procedia 6:541–549
23. Samadi M, Rakhtala SM (2019) Reducing cost and size in photovoltaic systems using three-level boost converter based on fuzzy logic controller. Iran J Sci Technol Trans Electr Eng 43:313–323
24. Farajdadian S, Hosseini SMH (2019) Design of an optimal fuzzy controller to obtain maximum power in solar power generation system. Solar Energy 161–178
25. Yaqin EN, Abdullah AG, Hakim DL, Nandiyanto ABD (2018) MPPT based on fuzzy logic controller for photovoltaic system using PSIM and Simulink. IOP Conf Ser: Mater Sci Eng 288(1), art. no. 012066
26. Youssef A-R (2015) Maximum power point tracking of a wind power system based on five phase PMSG using optimum torque control. In: 17th international middle-east power system conference (MEPCON'15), Mansoura University, Egypt, 15–17 Dec 2015
27. Algarin CR, Giraldo JT, Alvarez OR (2017) Fuzzy logic based MPPT controller for a PV system. Energies 10(12), art. no. 2036
28. Ray P, Sinha AK (2016) Modelling and simulation of grid power management with modified fuzzy logic based MPPT tracking with hybrid power energy system. In: 2015 international conference on energy, power and environment: towards sustainable growth, ICEPE 2015, art. no. 7510168
29. Lopez-Santos O, Garcia G, Martinez-Salamero L, Giral R, Vidal-Idiarte E, Merchan-Riveros MC, Moreno-Guzman Y (2019) Analysis, design, and implementation of a static conductance-based MPPT method. IEEE Trans Power Electron 34(2):1960–1979, art. no. 8359380
30. Nirudh J, Somchat J (2019) Estimation of solar potential for thailand using adaptive neurofuzzy inference system models. J Sol Energy Eng Trans ASME 141(6):061009
31. Amara K, Fekik A, Hocine D, Bakir ML, Bourennane E-B, Malek TA, Malek A (2018) Improved performance of a PV solar panel with adaptive neuro fuzzy inference system

ANFIS based MPPT. In: 7th international IEEE conference on renewable energy research and applications, ICRERA 2018, pp 1098–1101, art. no. 8566818

32. Pilakkat D, Kanthalakshmi S (2018) Drift free variable step size perturb and observe MPPT algorithm for photovoltaic systems under rapidly increasing insolation. Electronics 22(1): 19–26

33. Pilakkat D, Kanthalakshmi S (2019) An improved P&O algorithm integrated with artificial bee colony for photovoltaic systems under partial shading conditions. Sol Energy 37–47

34. Salman S, Ai X, Wu Z (2018) Design of a P-&-O algorithm based MPPT charge controller for a stand-alone 200 W PV system. Prot Control Mod Power Syst 3(1), art. no. 25

35. Jiang LL, Srivatsan R, Maskell DL (2018) Computational intelligence techniques for maximum power point tracking in PV systems: a review. Renew Sustain Energy Rev 85:14–45

36. Mohammedi A, Mezzai N, Rekioua D, Rekioua T (2014) Impact of shadow on the performances of a domestic photovoltaic pumping system incorporating an MPPT control: a case study in Bejaia, North Algeria. Energy Convers Manag 84:20–29

37. Farh HMH, Othman MF, Eltamaly AM (2018) Maximum power extraction from grid-connected PV system. In: 2017 Saudi Arabia smart grid conference, SASG 2017, pp 1–6

38. Eltamaly AM (2018) Performance of MPPT Techniques of Photovoltaic Systems Under Normal and Partial Shading Conditions. Adv Renew Energ Power Technol 1:115–161

39. Eltamaly AM, Farh HMH, Al Saud MS (2019) Impact of PSO reinitialization on the accuracy of dynamic global maximum power detection of variant partially shaded PV systems. Sustainability (Switzerland) 11(7), art. no. 2091

40. Smida MB, Sakly A, Vaidyanathan S, Azar AT (2018) Control-based maximum power point tracking for a grid-connected hybrid renewable energy system optimized by particle swarm optimization. Adv Syst Dyn Control 58–89

41. Priyadarshi N, Ramachandaramurthy VK, Padmanaban S, Azam F (2019) An ant colony optimized MPPT for standalone hybrid PV-wind power system with single Cuk converter. Energies 12(1), art. no. en12010167

42. Algarin CR, Alvarez OR, Castro AO (2018) Data from a photovoltaic system using fuzzy logic and the P&O algorithm under sudden changes in solar irradiance and operating temperature. Data Brief 21:1618–1621

43. Bataineh K, Eid N (2018) A hybrid maximum power point tracking method for photovoltaic systems for dynamic weather conditions. Resources 7(4):68

44. Soufi Y, Bechouat M, Kahla S (2017) Fuzzy-PSO controller design for maximum power point tracking in photovoltaic system. Int J Hydrogen Energy 42(13):8680–8688

45. Merchaoui M, Sakly A, Mimouni MF (2018) Particle swarm optimisation with adaptive mutation strategy for photovoltaic solar cell/module parameter extraction. Energy Convers Manag 175:151–163

46. Batarseh MG, Za'ter ME (2018) Hybrid maximum power point tracking techniques: a comparative survey, suggested classification and uninvestigated combinations. Sol Energy 169:535–555

47. Ahmad R, Murtaza AF, Sher HA (2019) Power tracking techniques for efficient operation of photovoltaic array in solar applications – a review. Renew Sustain Energy Rev 82–102

48. Rekioua D, Matagne E (2012) Optimization of photovoltaic power systems: modelization, simulation and control. Green energy and technology, vol 102

49. Boutabba T, Alibi A, Chrifi-Alaoui L, Ouriagli M, Drid, S, Mehdi D, Benbouzid MEH (2018) DSPACE real-time implementation sliding mode maximum power point tracker for photovoltaic system. In: 2018 7th international conference on systems and control, ICSC 2018, pp 137–141, 8587827

50. Yin L, Yu S, Zhang X, Tang Y (2017) Simple adaptive incremental conductance MPPT algorithm using improved control model. J Renew Sustain Energy 9:065501

51. Kennedy J, Eberhart R (1995) Particle swarm optimization. In: Proceedings of IEEE international conference on neural networks, Perth, vol IV, pp 1942–1948

52. Kamejima T, Phimmasone V, Kondo Y, Miyatake M (2011) The optimization of control parameters of PSO based MPPT for photovoltaics. In: IEEE PEDS 2011, Singapore, 5–8 Dec 2011
53. Suryavanshi R, Joshi DR, Jangamshetti SH (2012) PSO and P&O based MPPT technique SPV panel under varying atmospheric conditions. Int J Eng Technol (IJEIT) 1(3)
54. Kazmi SMR, Goto H, Guo H-J, Ichinokura O (2010) Review and critical analysis of the research papers published till date on maximum power point tracking in wind energy conversion system. In: 2010 IEEE energy conversion congress and exposition, Atlanta, GA, USA, 12–16 Sept 2010, pp 4076–4082
55. Colorni A, Dorigo M, Maniezzo V (1991) Distributed optimization by ant colonies. In: Conference: proceedings of ECAL91—European conference on artificial life, Jan 1991
56. Rajkumar RK, Ramachandaramurthy VK, Yong BL, Chia DB (2011) Techno-economical optimization of hybrid PV/wind/battery system using neuro-fuzzy. Energy 36(8):5148–5153. https://doi.org/10.1016/j.energy.2011.06.017
57. Tadjine K, Aissou S, Mebarki NE, Rekioua D, Logerais P-O (2018) Development of a LabVIEW interface to maximize the photovoltaic power under different outdoor conditions. In: Proceedings of 2017 international renewable and sustainable energy conference, IRSEC 2017, 8477251
58. Serir C, Rekioua D (2017) Contrôle vectoriel-logique floue pour le pompage photovoltaïque (Vector control of a motor pump system powers by a photovoltaic generator optimized by fuzzy logic control). Revue Roumaine des Sciences Techniques Serie Electrotechnique et Energetique 62(4):375–380
59. Abdullah MA, Yatim AHM, Tan CW, Saidur R (2012) A review of maximum power point tracking algorithms for wind energy systems. Renew Sustain Energy Rev 16(5):3220–3227. https://doi.org/10.1016/j.rser.2012.02.016
60. Eroglu M, Dursun E, Sevencan S, Song J, Yazici S, Kilic O (2011) A mobile renewable house using PV/wind/fuel cell hybrid power system. Int J Hydrogen Energy 36(13):7985–7992
61. Karakoulidis K, Mavridis K, Bandekas DV, Adoniadis P, Potolias C, Vordos N (2011) Techno-economic analysis of a stand-alone hybrid photovoltaic-diesel-battery-fuel cell power system. Renew Energy 36(8):2238–2244. https://doi.org/10.1016/j.renene.2010.12.003
62. Hosseinian H, Damghani H (2019) Ideal planning of a hybrid wind-PV-diesel microgrid framework with considerations for battery energy storage and uncertainty of renewable energy resources. In: 2019 IEEE 5th conference on knowledge based engineering and innovation, KBEI 2019, pp 911–916, art. no. 8734947
63. Qandil MD, Abbas AI, Qandil HD, Al-Haddad MR, Amano RS (2019) A stand-alone hybrid photovoltaic, fuel cell, and battery system: case studies in Jordan
64. Rekioua D (2014) Hybrid wind systems. Green energy and technology, pp 163–183. 9781447164241
65. Lu X, Qu Y, Wang Y, Qin C, Liu G (2018) A comprehensive review on hybrid power system for PEMFC-HEV: Issues and strategies. Energy Convers Manag 171:1273–1291
66. Assaf J, Shabani B (2018) Experimental study of a novel hybrid solar-thermal/PV-hydrogen system: towards 100% renewable heat and power supply to standalone applications. Energy 157:862–876
67. Jamshidi M, Askarzadeh A (2019) Techno-economic analysis and size optimization of an off-grid hybrid photovoltaic, fuel cell and diesel generator system. Sustain Cities Soc 44:310–320
68. Elhadidy MA, Shaahid SM (1998) Feasibility of hybrid (wind + solar) power systems for Dhahran Saudi Arabia. In: World renewable energy congress, Florence, Italy, 20–25 Sept 1998, vol 5

Chapter 4
Storage in Hybrid Renewable Energy Systems

4.1 Introduction

Energy storage is a dominant factor. It can reduce power fluctuations, enhance system flexibility and enable the storage and dispatch of electricity generated by variable renewable energy sources such as wind and solar. Different storage technologies are used with wind energy system or with hybrid wind systems. It can be electrical, chemical or electrochemical and mechanical or thermal (Fig. 4.1).

The following Fig. 4.2 gives a summary of the most used storage technologies. Capacitors are based on the physical separation of the electrical charge through a dielectric medium and the super-capacitors are based on the separation of chemically charged species at an electrified interface between a solid electrode and an electrolyte. But batteries and fuel cells are based on the conversion of chemical energy into electrical energy [1] (Table 4.1).

4.2 Electrochemical Storage

Batteries are normally required for most standalone applications. In renewable energy systems, common battery types used for storage are lead acid, Li-ion and hybrid flow batteries. Lead acid batteries are the most used due to their performances and are used in PV, wind turbine and hybrid systems for traction as in EV or for micro-grids and off-grid systems. Lithium-ion batteries are the best choice for high-power mobile applications (notebooks, laptops and mobile phones). But flow batteries are more capable of storing energy for longer durations than lithium-ion batteries (Table 4.2).

© Springer Nature Switzerland AG 2020
D. Rekioua, *Hybrid Renewable Energy Systems*, Green Energy and Technology,
https://doi.org/10.1007/978-3-030-34021-6_4

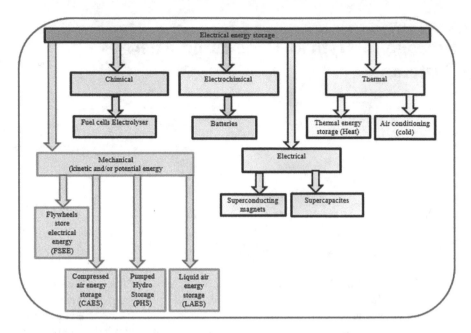

Fig. 4.1 Types of energy storage systems

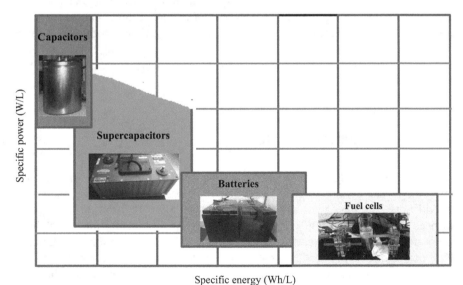

Fig. 4.2 Most used storage technologies

Table 4.1 Some important parameters of battery, FCs and SC [2–4]

	Batteries	FCs	SCs
Power range	Until 50 MW	Until 100 MW	Until 1 MW
Discharge duration	10 h	24 h	0.3 h
Power density	Less to 1 kW/kg	Less to 05 kW/kg	Up to 10 kW/kg
Energy density	Up to 100 Wh/kg	Up to 1000 Wh/kg	Less to 10 Wh/kg

Table 4.2 Most battery types

Batteries	Types		Proprieties
Lead acid	Flooded		– Cheap
	Valve regulated lead acid (VRLA)	Absorbed glass mat (AGM)	– Reliable
		Gel	
	Flow	Redox flow batteries (RFB)	
		Hybrid flow batteries (ZNBR)	
Lithium-ion (Li-ion)	Lithium cobalt oxide (LCO)		– Long cycle life
	Lithium manganese oxide (LMO)		– Light-weight
	Lithium nickel manganese cobalt oxide (NMC)		– No maintenance – High energy density
	Lithium nickel cobalt aluminum oxide (NCA)		– Longevity – Good performances
	Lithium iron phosphate (LFP)		
Nickel or nickel–cadmium (NiCd)			– Cheap and robust – Tolerability to extreme temperatures – Low maintenance – Needs ventilation
Nickel–zinc (Ni-Zn)			– Better environmental quality – High energy
Nickel–hydrogen			– Long cycle – Resistant to overcharge – Good energy density – High cost – High pressure cell – Low volumetric energy density
Nickel–metal hydride batteries			– High self-discharge – Failure leading to high pressure
Sodium			– Not explosive

(continued)

Table 4.2 (continued)

Batteries	Types	Proprieties
Sodium–sulfur (Nas)		– High energy density – Long cycle life – Use inexpensive materials – Fast response in charging and discharging
Sodium metal chloride		– Used for electric cars and in stationary systems – Relatively high energy density – No electric self-discharge

4.2.1 Application in Hybrid PV-Wind Turbine System

Batteries are associated to buck–boost converter (Fig. 4.3).

In a hybrid system, batteries are connected to the DC bus via DC/DC converter (Fig. 4.4).

4.2.2 Battery Models

In literature, there are different electrochemical battery models [5–66]. They are used to provide information on energy storage, state of charge and discharge, battery voltage and current waveform, … (Tables 4.3, 4.4, 4.5, 4.6, 4.7 and 4.8).

4.2.3 Application in Photovoltaic System

An application to a standalone load supplied by PV with batteries storage (Fig. 4.5).

The global system under MATLAB/Simulink is (Fig. 4.6).

Fig. 4.3 Battery storage with buck–boost converter

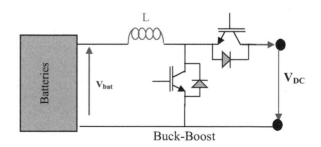

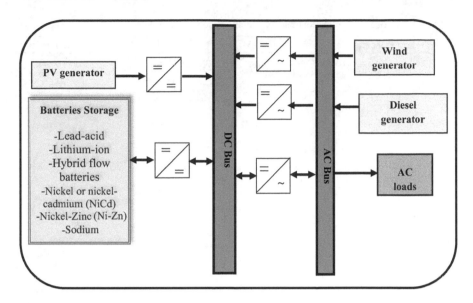

Fig. 4.4 Batteries storage in hybrid system

Table 4.3 Battery models based on equations

	Principle	Model	References
Peukert equation	An approach to the variation of a battery's capacity according to the rate of discharge	$C_1 = C_2 \cdot \left(\frac{I_{batt2}}{I_{batt1}}\right)^{n-1}$ $SOC(t_{k-1}) = 1 - \left(\frac{I_{batt}}{C}\right) \cdot t_{k-1}$ $\Delta SOC(t_k) = \frac{I_{batt2}}{C_1} \cdot \left(\frac{I_{batt2}}{I_{batt1}}\right)^{n-1} \cdot \Delta t$ $SOC(t_k) = SOC(t_{k-1}) + \Delta SOC(t_k)$	[7–9]
Shepherd equation	Describes the battery's electrochemical properties in terms of voltage and current.	$E_t = V_{oc} - R_{batt} \cdot I_{batt} - K_i(1 - Q)$	[10]
Unnewehr universal model	It is a simplified Shepherd equation	$E_t = E_0 - R_{batt} \cdot I_{batt} - K_i Q$ $I_{batt} = \frac{\pm E_{oc} - \sqrt{(E_{oc})^2 - 4 \cdot R_{batt} \cdot P}}{2 \cdot R_{batt}}$	[10, 11, 59]

The battery model is implemented under MATLAB/Simulink, as expressed in Fig. 4.7.

Some simulation results are represented in Fig. 4.8.

Table 4.4 An overview of simple battery models based on circuits

	Principle	Model	References
Improved simple model	This model varies the resistance which depends on the state of charge	$R_{batt} = \frac{R_0}{SOC}$ $SOC = 1 - \frac{I_{batt} \cdot t}{C_{10}}$	[12–16]
Modified battery model	In this model, the resistance of the battery varies as a function of its state of charge	$E_b = E_0 - k_{E_b} \cdot SOC$ $R_{batt} = R_0 - k_R \cdot SOC$	[12–18]

Table 4.5 An overview of battery Thevenin models

	Principle	Model	References
Thevenin model			[9, 10]
Resistive Thevenin model			[5]
Modified Thevenin equivalent model			[13]

Table 4.6 Linear and nonlinear battery models

	Principle	Model	References
Linear dynamic model		V_{batt}, R_p, R_1, C_1, R_2, C_2, R_3, C_3, C_0, E_0	[64]
Nonlinear dynamic model	The model considers the variation of the different variables with the battery state of charge, temperature and discharge rate.	C_1, I_{batt}, $R_c(SOC,T)$, $R_d(SOC,T)$, V_{batt}, $V_{oc}(SOC, T)$	[65]

Table 4.7 Dynamic battery models

	Principle	Model	References
Dynamic model	It is an empirical mathematical model	$V_{batt} = V_{oc} - R_{batt} \cdot I_{batt} - \dfrac{K}{SOC} I_{batt}$ — I_m, I_p, R_0, I_{batt}, Z_m, Z_p, V_{batt}, E_m, E_p	[66]
Dynamic model of third order		I_1, C, I_m, R_1, R_2, I_p, R, I_{batt}, E_m, $I_p(V_p)$, V_{batt}	[67]
Dynamic model of fourth order	It is an accurate and sophisticated model	C_d, C, R_W, R_d, R_p, I_p, I_s, R_s, I_{batt}, E_b, E_s, V_{batt}	[67]

Table 4.8 Other used battery models

	Principle	Model	References
Model from experimental tests	This model is based on a series of experimental tests that are carried out by investigating the graphical plots of the test data and the company's specifications	$C_p \cdot \dfrac{dv_p}{dt} = I_{batt} - \dfrac{v_p}{R_p}$ $C_1 \cdot \dfrac{dv_p}{dt} = I_{batt} - \dfrac{v_p}{R'}$ $V_{batt} = V_{oc} - R_c \cdot I_{batt} - V_{C_1}$	[13]
Model traction	It is used in electric and hybrid vehicles	$V_{batt} = E_b - R_{batt} \cdot I_{batt} - K \int \left(\dfrac{I_{batt}}{Q}\right) dt$ $\dfrac{C_{batt}}{C_{10}} = \dfrac{1.67}{1 + 0.67 \left(\dfrac{I}{I_{10}}\right)^{0.9}}(1 + 0.005\Delta T)$ $SOC = 1 - \dfrac{Q}{C_{batt}}$ $Q = I_{batt} \cdot t$	[14, 16–28]

(continued)

Table 4.8 (continued)

Principle	Model	References
	$V_{\text{batt}-\text{charge}} = n_b[2. + 0.16.\text{SOC}]$ $+ n_b \cdot \frac{I_{\text{batt}}}{C_{10}} \left[\frac{6}{1+I_{\text{batt}}^{1.3}} - \frac{0.27}{\text{SOC}^{1.5}} + 0.002 \right] \cdot (1 - 0.007.\Delta T)$ $V_{\text{batt}-\text{discharge}} = n_b[1.965. + 0.12.\text{SOC}]$ $-n_b \cdot \frac{I_{\text{batt}}}{C_{10}} \left[\frac{4}{1+I_{\text{batt}}^{1.3}} + \frac{0.27}{\text{SOC}^{1.5}} + 0.002 \right] \cdot (1 - 0.007.\Delta T)$	

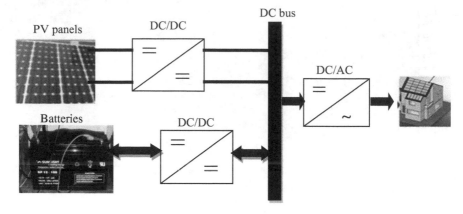

Fig. 4.5 PV system with batteries supplying a load

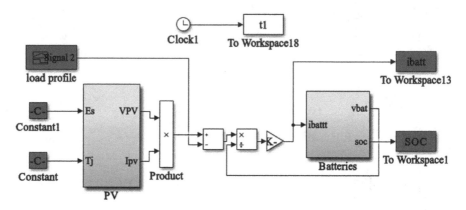

Fig. 4.6 PV system with battery storage

4.3 Mechanical Storage

4.3.1 Flywheel Electric Energy Storage (FEES)

Flywheel electric energy storage system includes a cylinder with a shaft connected to an electrical generator. Electric energy is converted by the generator to kinetic energy which is stored by increasing the flywheel's rotational speed. The stored energy is converted to electric energy via the generator, slowing the flywheel's rotational speed. For wind standalone applications, storage cost still represents a major economic restraint. Energy storage in wind systems can be achieved in different ways. However, the inertial energy storage adapts well to sudden power changes of the wind generator. Moreover, it allows obtaining very interesting power-to-weight characteristic in storing and delivering power. The reference speed for the flywheel is determined by [28].

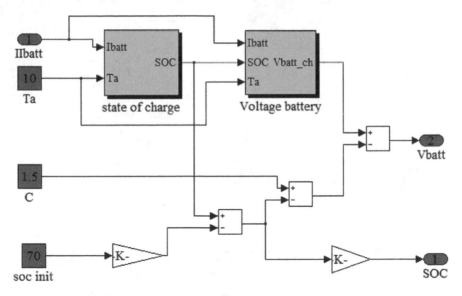

Fig. 4.7 Battery model in MATLAB/Simulink

4.3.2 Application in Wind System

In wind energy conversion system (WECS), flywheel energy storage (FES) is able to suppress fast wind power fluctuations. In this work, a WECS based on induction generator is simulated. The system is constituted of a wind turbine, an induction generator, a rectifier/inverter and a flywheel energy storage system (Fig. 4.9). The goal of the device is to provide a constant power and voltage to the load connected to the rectifier/inverter even if the speed varies. This can be achieved mainly by keeping the DC bus voltage at a constant value. The flywheel energy storage system contributes to maintain the delivered power to the load constant, as long as the wind power is sufficient [28].

To control the speed of the flywheel energy storage system, it is mandatory to find a reference speed which ensures that the system transfers the required energy by the load at any time. The reference speed can be determined by the reference energy. The power assessment of the overall system is given by [16, 28]:

$$P_{\text{ref}} = P_{\text{load}} - P_{\text{wind}} - \Delta P \tag{9}$$

where P_{ref} is the reference power, P_{load} is the load power, P_{wind} is the wind power and ΔP is the power required to maintain the DC voltage V_{dc} at a constant value.

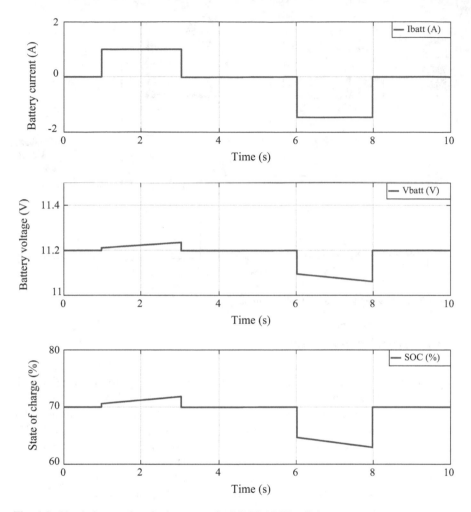

Fig. 4.8 Simulation results of a battery under MATLAB/Simulink

The system is simulated using MATLAB/Simulink software package with a wind power profile which provides the power continuously required by the load (Fig. 4.10).

The energy reference for the flywheel is:

$$E_{c\,ref} = E_c^{t1} + \int\limits_{t1}^{t2} P_{ref}.dt \tag{10}$$

With: E_c^{t1} is the initial energy.

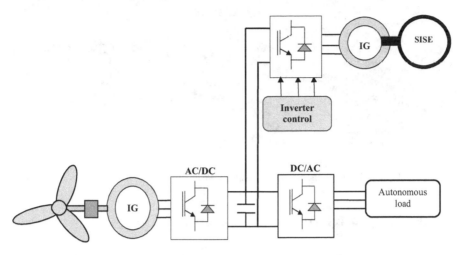

Fig. 4.9 FEES with IG in wind turbine system

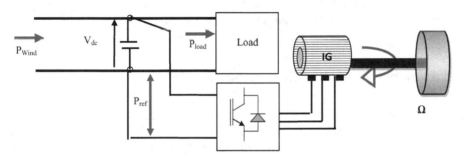

Fig. 4.10 WCES based on an induction generator with flywheel storage

The speed reference will be:

$$\Omega_{v\,\text{ref}} = \sqrt{\frac{2.E_{c\,\text{ref}}}{J_t}} \qquad (11)$$

$$J_t - J_{\text{IG}} + J_{\text{Flywheel}} \qquad (12)$$

The block diagram under MATLAB/Simulink is represented in Fig. 4.11.

Some simulation results are presented. It has been noticed that the stored fly-wheel energy depends on the available wind power and the required power by the load (Figs. 4.11 and 4.12).

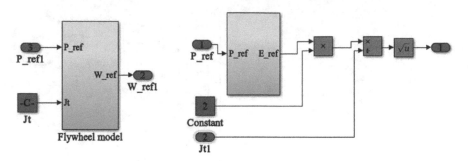

Fig. 4.11 Flywheel storage model under MATLAB/Simulink

4.3.3 Pumped Hydro Energy Storage (PHES)

Pumped hydro energy storage (PHES) system consists of a pumped hydro system with two large water reservoirs (upper and lower), an electric machine (motor/generator) and a reversible pump–turbine group. It is considered as an attractive alternative for energy storage due to its main advantages.

4.4 Super-Capacitor Energy Storage (SES)

It is known as electric double-layer capacitors, as super-capacitors (SC), electro-chemical double-layer capacitors (EDLCs) or ultra-capacitors. They use polarized liquid layers between conducting ionic electrolyte and conducting electrode to increase the capacitance. They allow a much higher energy density, with a high-power density, but the voltage varies with the energy stored. In wind energy conversion system, SES are used to suppress fast wind power fluctuations but at a small time scale. Thus, they can be considered only as a support for wind turbines systems and are generally combined with a battery system in a hybrid storage system [25] (Fig. 4.13).

The model in MATLAB/Simulink is as follow (Fig. 4.14).

4.5 Fuel Cells

There are six most commonly used types of fuel cells that are [20, 22]:

- Proton exchange polymer membranes fuel cells (PEMFC),
- Direct methanol fuel cell (DMFC),
- Phosphoric acid fuel cells (PAFC),
- Alkaline fuel cells (AFC),

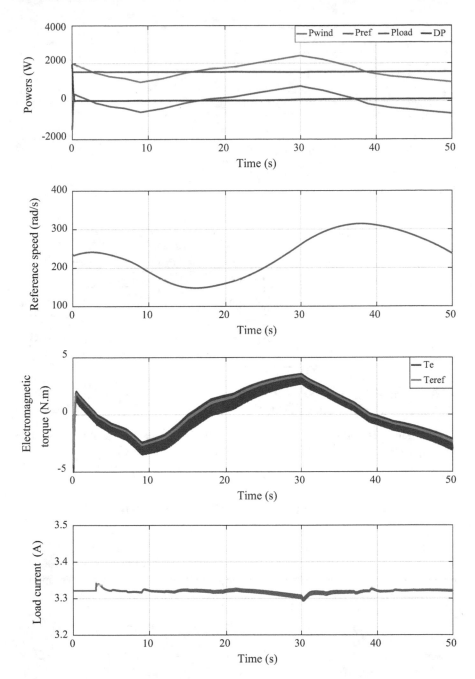

Fig. 4.12 Simulation results using flywheel in wind turbine system

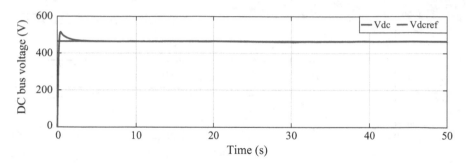

Fig. 4.12 (continued)

Fig. 4.13 SCs with buck–boost converter

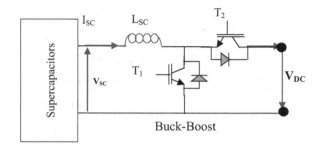

- Solid oxide fuel cells (SOFC),
- Molten carbonate fuel cells (MCFC).

The transport application is dominated by polymer electrolyte fuel cells, since they operate at a low temperature and have a very fast start time.

4.6   Electrical Characteristics of PEM Fuel Cells

The theoretical thermodynamic potential of the PEMFC fuel cell at 25 °C and 1 atm is about 1.23 V, but the actual potential (VPEMFC) of the cell decreases with respect to the steady-state thermodynamic potential when the current flows, and this deviation from the Nernst potential value [22, 28]:

$$V_{\text{PEMFC}} = E_{\text{Nernst}} - V_{\text{act}} - V_{\text{ohm}} - V_{\text{conc}} \qquad (13)$$

where

$$E_{\text{Nernst}} = 1.229 - 0.85 \times 10^{-3} \times (T - 298.15) + 4.31 \times 10^{-5}$$
$$\times \left[\ln\left(P_{\text{H}_2}^*\right) + \right] \frac{1}{2} \ln\left(P_{\text{O}_2}^*\right) \qquad (14)$$

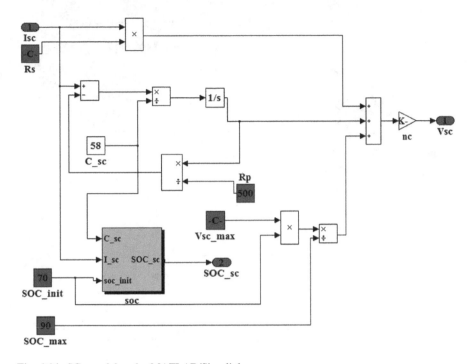

Fig. 4.14 SCs model under MATLAB/Simulink

$$V_{\text{act}} = \varepsilon_1 + \varepsilon_2 \times T + \varepsilon_3 \times T \times \ln\left(C_{O_2}^*\right) + \varepsilon_4 \times T \times \ln(I_{\text{PEMFC}}) \qquad (15)$$

$$V_{\text{ohm}} = \frac{I_{\text{PEMFC}}}{A} \left[\frac{1816 \times \left[1 + 0.03 \times \left(\frac{I_{\text{PEMFC}}}{A}\right) + 0.062 \times \left(\frac{T}{303}\right)^2 \times \left(\frac{I_{\text{PEMFC}}}{A}\right)^{2.5}\right]}{\left[\delta_{\text{H}_2\text{O}}/\text{SO}_3^- - 3 \times \left(\frac{I_{\text{PEMFC}}}{A}\right)\right] \times \exp\left[4.18 \times \left(\frac{T-303}{T}\right)\right]} \times l + A \times R_c \right]$$

$$(16)$$

$$V_{\text{conc}} = -B\left(1 - \frac{J}{J_{\text{max}}}\right) \qquad (14)$$

with (Fig. 4.15):

$$C_{O_2}^* = \frac{P_{O_2}^*}{\left(5.08 \times 10^6 \times e^{-\left(\frac{498}{T}\right)}\right)} \qquad (17)$$

$$P_{O_2}^* = P_{\text{cath}} \times \left(1 - \chi_{\text{H}_2\text{O}}^{\text{Sat}}\right) \qquad (18)$$

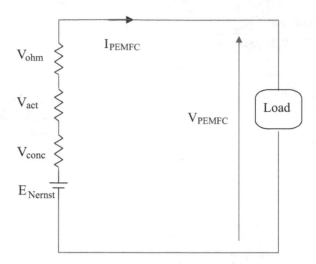

Fig. 4.15 Equivalent electrical circuit diagram of a PEMFC

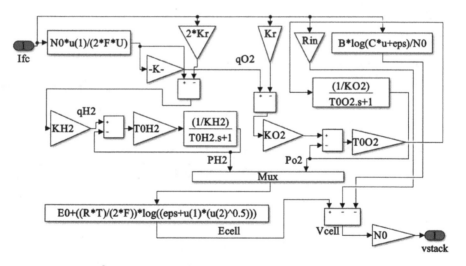

Fig. 4.16 Block diagram of PEMFCs model

In MATLAB/Simulink, the block diagram of PEMFC model is as follows (Fig. 4.16).

With (Fig. 4.17)

$$V_{PEMFC} = N_{cell} \times V_{cell} \qquad (19)$$

Fuel cells are connected to a load or a DC bus via a boost converter (Fig. 4.18).

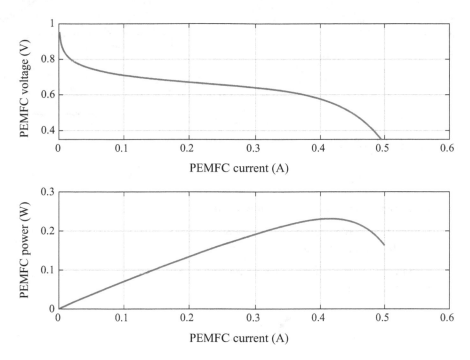

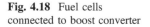

Fig. 4.17 Fuel cells voltage and power

Fig. 4.18 Fuel cells
connected to boost converter

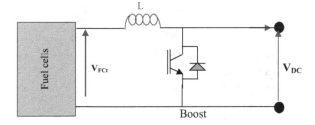

4.7   Super-Capacitors

The super-capacitor model consists of an ideal capacitor connected in series with a resistance (Fig. 4.19). Models with several RC branches and capacitance values as a function of the SC voltage are also proposed in the literature [26–28].

The voltage across a super-capacitor is given by:

$$P_{SC} = V_{SC} \times I_{SC} \tag{20}$$

Fig. 4.19 Equivalent electrical circuit of SCs

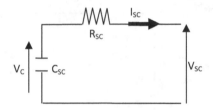

Thus, the power is:

$$V_{SC} = V_C - R_{SC} \times I_{SC} \tag{21}$$

The convention used and the following one:

$P_{sc} > 0$ in traction (discharge of the super-capacitor)
$P_{sc} < 0$ in braking (super-capacitor charge).

The super-capacitor behaves like a current integrator. The no-load voltage of the SC can be calculated as follows:

$$V_C = V_C(0) - \frac{1}{C} \int_0^t I_{SC}(u).du \tag{22}$$

The energy contained in the SC is a function of its no-load voltage. Therefore, the maximum energy is obtained for the maximum no-load voltage.

$$E_{SC} = \frac{1}{2} C_{SC} \times V_C \tag{23}$$

The state of charge is the ratio between the energy contained in the SC and the maximum energy.

$$SOC = \frac{E_{sc}(t)}{E_{scmax}} \tag{24}$$

To satisfy the power and energy demand of a given application, an assembly of several SCs is used. When the number of SC (N_{sc}) are connected in series, the voltage of the SC pack and the equivalent capacity are given by the following relationships:

$$U_{SC} = V_{SC} \times N_{SC} \tag{25}$$

$$Ceq_{CS} = \frac{C_{SC}}{N_{SC}} \tag{26}$$

The model of the SC under MATLAB/Simulink is as follows (Fig. 4.20). Voltage and current variations are given (Fig. 4.21).

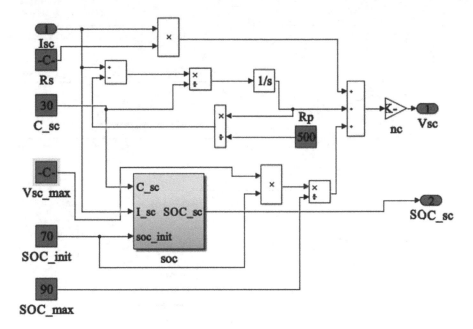

Fig. 4.20 SCs model under MATLAB/Simulink

4.8 Hybrid Storage

4.8.1 FC/SC Hybrid Storage

The energy flow is provided by the two DC/DC converters. The DC/DC boost converter allows a single direction of the energy flow, which is from PEMFC to the DC bus, and kept the voltage to a constant value. The energy flows in both directions via the buck–boost DC/DC converter between the SC and the DC bus (Figs. 4.22 and 4.23).

4.8.2 Application 1: FCS/SCs for EVs

Figure 4.24 provides the structure of the proposed hybrid power system. The complete architecture is prepared in DC bus. Fuel cells (FCs) and SCs are the subsystems of DC bus. The load is an electric vehicle system, the power transfer happens between these components via an energy management, which is based on control energy flows by controlling the DC/DC converters (Fig. 4.24). The SC/

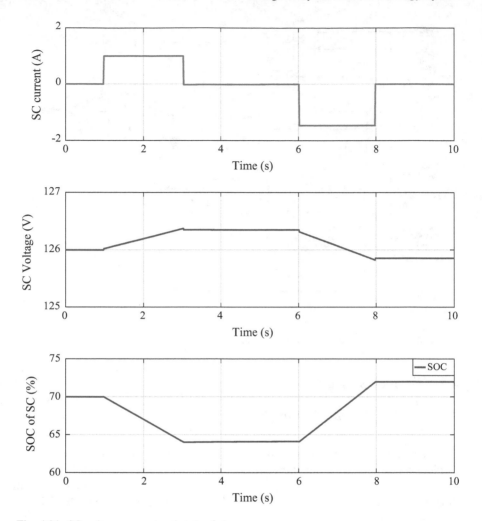

Fig. 4.21 SC voltage, current and state of charge

Buck–Boost converter assembly is connected to the DC bus via inductor L_s, its function is to ensure that the current does not exceed the limit value, and not to allow the SC to reach the voltage of the DC bus [14, 18–22].

An application has been made on an electric vehicle (Fig. 4.25).

4.8.3 Fuel Cell/Battery System

The model in MATLAB/Simulink is follows (Fig. 4.26).

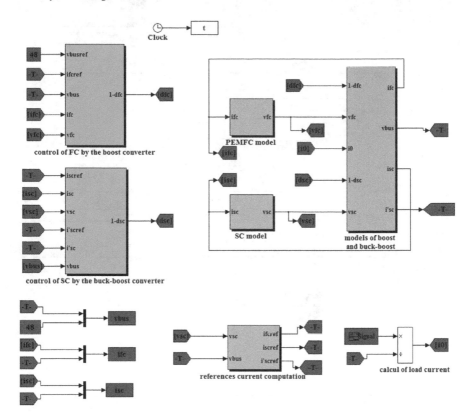

Fig. 4.22 Hybrid FC/SC storage in MATLAB/Simulink

4.8.4 Application 2: FCs/Batteries for EVs

An application to EV is made. The block diagram can be presented as (Fig. 4.27). The model in MATLAB/Simulink is (Fig. 4.28).

4.8.5 Application to Wind Turbine System

The wind turbine system is considered to be the main source. The battery and fuel cells are considered as storage systems. The electrolyzer is considered as an auxiliary charge to dissipate excess production. In this system, a control strategy is necessary to protect batteries and charge them with the excess wind power (Fig. 4.29).

The algorithm will be as follows (Fig. 4.30).

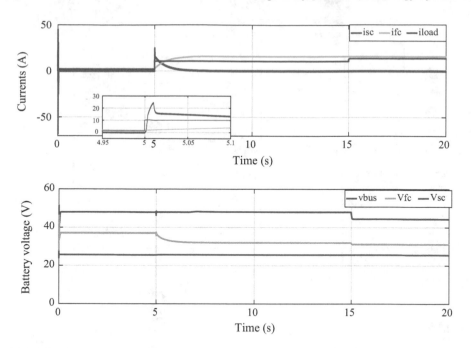

Fig. 4.23 Simulation results of hybrid FC/CS storage

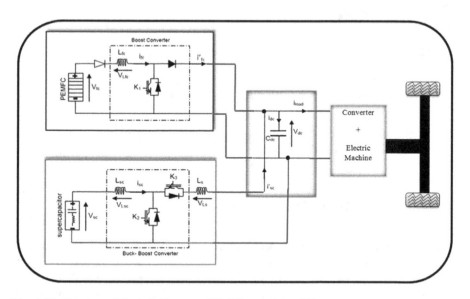

Fig. 4.24 Structure of the hybrid storage SCs/FCs supplying EV system

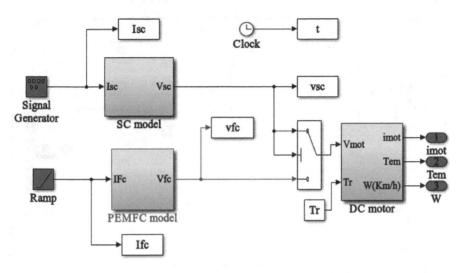

Fig. 4.25 Hybrid storage SCs/FCs supplying DC motor

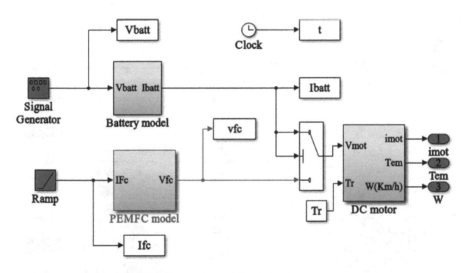

Fig. 4.26 Hybrid storage battery/FCs supplying DC motor

- when the battery's state of charge is at its maximum (SOC = SOC$_{max}$), the electrolyzer is activated to produce hydrogen, which is delivered to the hydrogen storage tanks.
- when there is a power deficit ($P_{load} < 0$), the batteries produce the needed power.

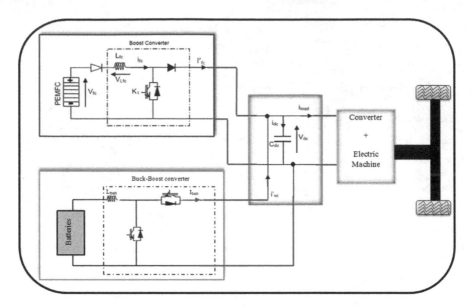

Fig. 4.27 Fuel cell/batteries system for EVs

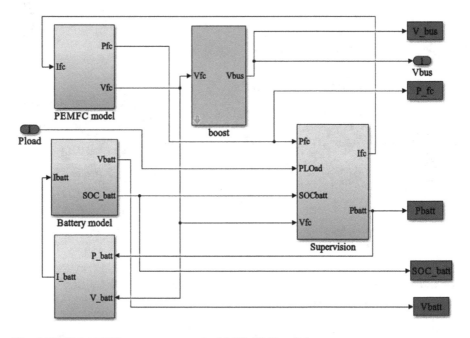

Fig. 4.28 Hybrid FC/battery storage under MATLAB/Simulink

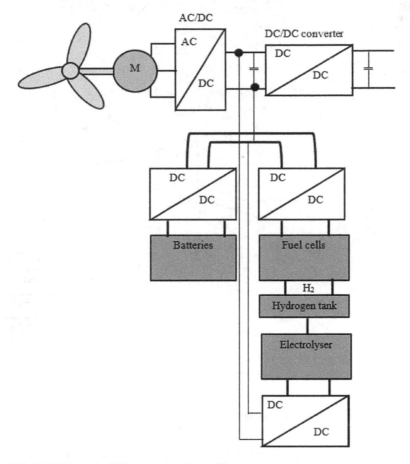

Fig. 4.29 Hybrid batteries/FC storage in wind turbine system

- if the SOC of the battery is lower than the minimum SOC value (SOC = SOC$_{min}$), the FCs start producing the power necessary for charging by using hydrogen stored through storage. The residual power is supplied by fuel cells when there is a power deficit and consumed by the electrolyzer when there is a power excess.

The power balance is:

$$P_{load} = P_{wind} + P_{batt} + P_{FCs}. \tag{27}$$

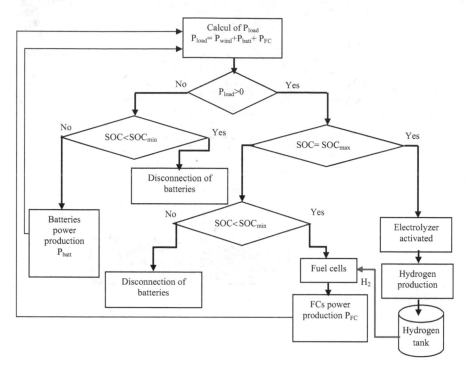

Fig. 4.30 Algorithm of wind turbine with batteries and fuel cells

4.8.6   Super-Capacitors/Batteries System

It is difficult to recommend the use of batteries or super-capacitors, as the two technologies have very different characteristics. The use of super-capacitors is generally preferred because of their energy efficiency and specific power, but their low capacity can be a drawback. On the other hand, batteries are capable of storing a large amount of energy but are penalized by their specific power. One solution is to combine batteries and super-capacitors to cumulate their advantages (power and energy), but this can increase the complexity and cost. The model under MATLAB/ Simulink can be represented as (Fig. 4.31).

4.8.7   Application 3: SCs/Batteries for EVs

The block diagram of SCs/batteries supplying EV can be given as (Fig. 4.32).

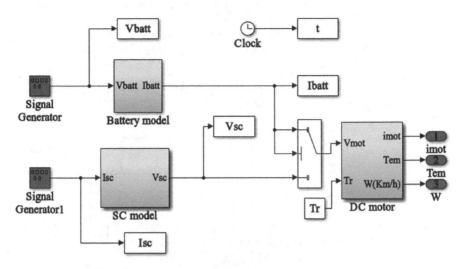

Fig. 4.31 Hybrid battery/SC storage supplying DC motor

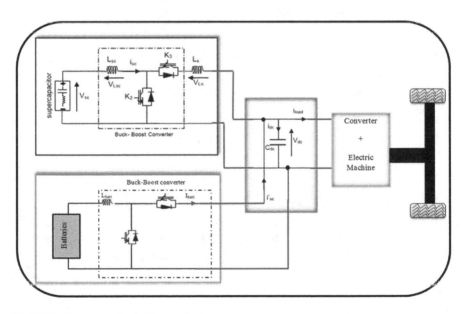

Fig. 4.32 Super-capacitor/battery system

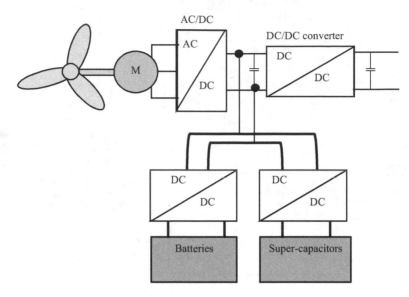

Fig. 4.33 Hybrid storage in wind turbine system

4.8.8 Application in Wind Turbine System

Hybrid batteries/SCs storage can be applied to wind turbine system (Fig. 4.33). The power balance is:

$$P_{\text{load}} = P_{\text{wind}} + P_{\text{batt}} + P_{\text{SC}}. \tag{28}$$

4.8.9 Multiple Storage

A multi-storage combined FCs, batteries and super-capacitors have been used for EVs. A representation of this system can be shown in Fig. 4.34.

4.9 Conclusions

An overview of the most used storage has been presented with different models implemented in MATLAB/Simulink. For hybrid storage, a power management control is necessary which is explained in Chap. 6.

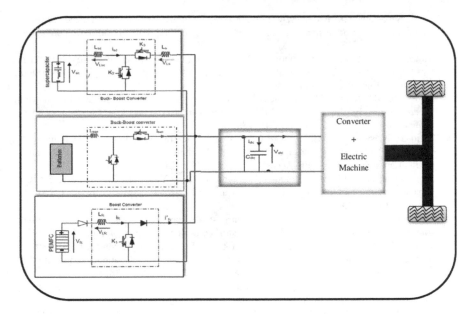

Fig. 4.34 FC/SC/battery storage system

References

1. https://www.cap-xx.com/resource/energy-storage-technologies/
2. Krishan O, Suhag S (2018) An updated review of energy storage systems: classification and applications in distributed generation power systems incorporating renewable energy resources. Int J Energy Res (October):1–40
3. Gur TM (2018) Review of electrical energy storage technologies, materials and systems: challenges and prospects for large-scale grid storage. Energy Environ Sci 11(10):2696–2767
4. Zakeri B, Syri S (2015) Electrical energy storage systems: a comparative life cycle cost analysis. Renew Sustain Energy Rev 42:569–596
5. Rekioua D (2014) Modeling of storage systems (book chapter). Green Energy Technol, 107–131, 9781447164241
6. http://Www.Rollsroyce.Com/Marine/Products/Electrical_Power_Systems/Storage/Index.Jsp
7. Noorhaninah S, Missman B (2010) Simulation design of a stand-alone photovoltaic (PV) inverter, a report submitted in partial fulfilment of the requirements for the award of the degree of bachelor of electrical engineering. Faculty of Electrical Engineering, Universiti Teknologi Malaysia
8. Dûrr M (2006) Dynamic model. J Power Sources 161:1400–1411
9. Zoroofi S (2008) Modeling and simulation of vehicular power systems. Thesis of master. University of Technologie, Chalmers
10. Ziyad M, Margaret A, William A (1992) A mathematical model for lead acid battery. IEEE Trans Energy Convers 7(1)
11. Gergaud O, Robin G, Multon B, Ahmed HB (2003) Energy modeling of a lead-acid battery within hybrid wind/photovoltaic systems. In: European power electronic conference, pp 1–8
12. Achaibou N, Haddadi M, Malek A (2008) Lead acid batteries simulation including experimental validation. J Power Sources 185:1484–1491
13. Francisco M, Longatt G (2006) Circuit based battery models: a review. In: 2do Congreso Iberoamericano De Estudiantes De Ingeniería Eléctrica (II CIBELEC 2006), pp 1–5

14. Zhang W, Maleki A, Rosen MA, Liu J (2018) Optimization with a simulated annealing algorithm of a hybrid system for renewable energy including battery and hydrogen storage. Energy 163:191–207
15. Mohammedi A, Rekioua D, Rekioua T, Mebarki NE (2018) Comparative assessment for the feasibility of storage bank in small scale power photovoltaic pumping system for building application. Energy Convers Manag 172:579–587
16. Amrouche SO, Rekioua D, Rekioua T (2016) Overview of energy storage in renewable energy systems. In: Proceedings of 2015 IEEE international renewable and sustainable energy conference, IRSEC 2015. Art. no. 7454988
17. Tariq M, Maswood AI, Gajanayake CJ, Gupta AK (2018) Modeling and integration of a lithium-ion battery energy storage system with the more electric aircraft 270 v DC power distribution architecture. IEEE Access 6:41785–41802. Art. no. 8438873
18. Mohammedi A, Rekioua D, Rekioua T, Bacha S (2016) Valve regulated lead acid battery behavior in a renewable energy system under an ideal Mediterranean climate. Int J Hydrogen Energy 41(45):20928–20938
19. Fathabadi H (2018) Novel high-efficient large-scale stand-alone solar/wind hybrid power source equipped with battery bank used as storage device. J Energy Storage 17:485–495
20. Tazerart F, Mokrani Z, Rekioua D, Rekioua T (2015) Direct torque control implementation with losses minimization of induction motor for electric vehicle applications with high operating life of the battery. Int J Hydrogen Energy 40(39):13827–13838
21. Jiang Y, Kang L, Liu Y (2019) A unified model to optimize configuration of battery energy storage systems with multiple types of batteries. Energy, 552–560
22. Amrouche SO, Rekioua D, Rekioua T, Bacha S (2016) Overview of energy storage in renewable energy systems. Int J Hydrogen Energy 41(45):20914–20927
23. Ghanaatian M, Lotfifard S (2019) Control of flywheel energy storage systems in the presence of uncertainties. IEEE Trans Sustain Energy 10(1):36–45. Art. no. 8329549
24. Hajiaghasi S, Salemnia A, Hamzeh M (2019) Hybrid energy storage system for microgrids applications: a review. J Energy Storage 21:543–570
25. Argyrou MC, Christodoulides P, Marouchos CC, Kalogirou SA (2018) Hybrid battery-supercapacitor mathematical modeling for PV application using Matlab/simulink. In: Proceedings 2018 53rd international universities power engineering conference, UPEC 2018, 8541933
26. Bracco S, Delfino F, Trucco A, Zin S (2018) Electrical storage systems based on sodium/ nickel chloride batteries: a mathematical model for the cell electrical parameter evaluation validated on a real smart microgrid application. J Power Sources 399:372–382
27. Mohamad F, Teh J, Lai C-M, Chen L-R (2018) Development of energy storage systems for power network reliability: a review. Energies 11(9). Art. no. 2278
28. Mebarki N, Rekioua T, Mokrani Z, Rekioua D, Bacha S (2016) PEM fuel cell/battery storage system supplying electric vehicle. Int J Hydrogen Energy 41(45):20993–21005
29. Karaoglan MU, Kuralay NS, Colpan CO (2019) Investigation of the effects of battery types and power management algorithms on drive cycle simulation for a range-extended electric vehicle powertrain. Int J Green Energy 16(1):1–11
30. Fang H, Zhang J, Gao J (2010) Optimal operation of multi-storage tank multi-source system based on storage policy. J Zhejiang Univ Sci A Appl Phys Eng 11(8):571–579
31. Becker J, Schaeper C, Sauer DU (2012) Energy management system for a multi-source storage system electric vehicle. In: 2012 IEEE vehicle power and propulsion conference (VPPC). Seoul, South Korea
32. Merei G, Berger C, Sauer DU (2013) Optimization of an off-grid hybrid PV-wind-diesel system with different battery technologies using genetic algorithm. Sol Energy 97:460–473. https://doi.org/10.1016/j.solener.2013.08.016
33. Mendis N, Muttaqi KM, Perera S (2014) Management of battery-supercapacitor hybrid energy storage and synchronous condenser for isolated operation of PMSG based variable-speed wind turbine generating systems. IEEE Trans Smart Grid 5(2):944–953. Art. no. 6720212. https://doi.org/10.1109/tsg.2013.2287874

34. Krishan O, Suhag S (2018) An updated review of energy storage systems: Classification and applications in distributed generation power systems incorporating renewable energy resources. Int J Energy Res. http://onlinelibrary.wiley.com/journal/10.1002/(ISSN)1099-114X, https://doi.org/10.1002/er.4285

35. Bopp G, Gabler H, Preiser K, Sauer DU, Schmidt H (1998) Energy storage in photovoltaic stand-alone energy supply systems. Prog Photovolt Res Appl 6(4):271–291

36. Jossen A, Garche J, Sauer DU (2004) Operation conditions of batteries in PV applications. Solar Energy 76(6):759–769. https://doi.org/10.1016/j.solener.2003.12.013

37. Akinyele DO, Rayudu RK (2014) Review of energy storage technologies for sustainable power networks. Sustain Energy Technol Assess 8:74–91. http://www.journals.elsevier.com/sustainable-energy-technologies-and-assessments, https://doi.org/10.1016/j.seta.2014.07.004

38. Chen H, Cong TN, Yang W, Tan C, Li Y, Ding Y (2009) Progress in electrical energy storage system: a critical review. Prog Nat Sci 19(3):291–312. http://www.sciencedirect.com/science/journal/aip/10020071, https://doi.org/10.1016/j.pnsc.2008.07.014

39. Dunn B, Kamath H, Tarascon J-M (2011) Electrical energy storage for the grid: a battery of choices. Science 334(6058):928–935. http://www.sciencemag.org/content/334/6058/928.full.pdf, https://doi.org/10.1126/science.1212741

40. Abraham KM (2015) Prospects and limits of energy storage in batteries. J Phys Chem Lett 6 (5):830–844. http://pubs.acs.org/journal/jpclcd, https://doi.org/10.1021/jz5026273

41. Ibrahim H, Ilinca A, Perron J (2008) Energy storage systems-characteristics and comparisons. Renew Sustain Energy Rev 12(5):1221–1250. https://doi.org/10.1016/j.rser.2007.01.023

42. Luo X, Wang J, Dooner M, Clarke J (2015) Overview of current development in electrical energy storage technologies and the application potential in power system operation. Appl Energy 137:511–536. http://www.elsevier.com/inca/publications/store/4/0/5/8/9/1/index.htt

43. Gröger O, Gasteiger HA, Suchsland J-P (2015) Review-electromobility: batteries or fuel cells? J Electrochem Soc 162(14):A2605–A2622. http://jes.ecsdl.org/content/by/year, https://doi.org/10.1149/2.0211514jes

44. Burke AF (2007) Batteries and ultracapacitors for electric, hybrid, and fuel cell vehicles. Proc IEEE 95(4):806–820. Art. no. 4168012. https://doi.org/10.1109/jproc.2007.892490

45. Lukic SM, Cao J, Bansal RC, Rodriguez F, Emadi A (2008) Energy storage systems for automotive applications. IEEE Trans Ind Electron 55(6):2258–2267. https://doi.org/10.1109/tie.2008.918390

46. Zhang L, Hu X, Wang Z, Sun F, Deng J, Dorrell DG (2018) Multiobjective optimal sizing of hybrid energy storage system for electric vehicles. IEEE Trans Veh Technol 67(2):1027–1035. http://ieeexplore.ieee.org/xpl/tocresult.jsp?isnumber=8039128&punumber=25, https://doi.org/10.1109/tvt.2017.2762368

47. Schupbach RM, Balda JC, Zolot M, Kramer B (2003) Design methodology of a combined battery-ultracapacitor energy storage unit for vehicle power management, In: PESC record—IEEE annual power electronics specialists conference, vol 1, pp 88–93

48. Emadi A, Rajashekara K, Williamson SS, Lukic SM (2005) Topological overview of hybrid electric and fuel cell vehicular power system architectures and configurations. IEEE Trans Veh Technol 54(3):763–770. https://doi.org/10.1109/tvt.2005.847445

49. Cao J, Emadi A (2012) A new battery/ultracapacitor hybrid energy storage system for electric, hybrid, and plug-in hybrid electric vehicles. IEEE Trans Power Electron 27(1):122–132. Art. no. 5764539. https://doi.org/10.1109/tpel.2011.2151206

50. Lukic SM, Wirasingha SG, Rodriguez F, Cao J, Emadi A (2006) Power management of an ultracapacitor/battery hybrid energy storage system in an HEV. In: 2006 IEEE vehicle power and propulsion conference, VPPC 2006. Art. no. 4211267. ISBN: 1424401585; 978-142440158-1. https://doi.org/10.1109/vppc.2006.364357

51. Dorn-Gomba L, Chemali E, Emadi A (2018) A novel hybrid energy storage system using the multi-source inverter. In: Conference proceedings—IEEE applied power electronics conference and exposition—APEC, pp 684–691. ISBN: 978-153861180-7. https://doi.org/10.1109/apec.2018.8341086

52. Li X, Xu L, Hua J, Li J, Ouyang M (2008) Control algorithm of fuel cell/battery hybrid vehicular power system. In: 2008 IEEE vehicle power and propulsion conference, VPPC 2008. Art. no. 4677613. ISBN: 978-142441849-7. https://doi.org/10.1109/vppc.2008.4677613

53. Miñambres-Marcos VM, Guerrero-Martínez MÁ, Barrero-González F, Milanés-Montero MI (2017) A grid connected photovoltaic inverter with battery-supercapacitor hybrid energy storage. Sensors (Switzerland) 17(8). Art. no. 1856. http://www.mdpi.com/1424–8220/17/8/1856/pdf, https://doi.org/10.3390/s17081856

54. Hemmati R (2018) Optimal design and operation of energy storage systems and generators in the network installed with wind turbines considering practical characteristics of storage units as design variable. J Clean Prod 185:680–693

55. Song Z, Hofmann H, Li J, Hou J, Han X, Ouyang M (2014) Energy management strategies comparison for electric vehicles with hybrid energy storage system. Appl Energy 134:321–331

56. Rivera-Rodríguez EP (2019) Analysis of a lead-acid battery storage system connected to the DC bus of a four quadrants converter to a microgrid. Renew Energy Power Qual J 17:151–154

57. Argyrou MC, Christodoulides P, Kalogirou SA (2018) Energy storage for electricity generation and related processes: technologies appraisal and grid scale applications. Renew Sustain Energy Rev 94:804–821

58. Mutarraf MU, Terriche Y, Niazi KAK, Vasquez JC, Guerrero JM (2018) Energy storage systems for shipboard microgrids—a review. Energies 11(12). Art. no. 3492

59. Abdelli R, Rekioua D, Rekioua T, Bouzida A, Tounzi AM (2018) Control of the grid-side converter in wind conversion systems with flywheel energy storage and constant switching frequency. In: Proceedings of 2017 international renewable and sustainable energy conference, IRSEC 2017, 8477322

60. Guo F, Ye Y, Sharma R (2015) A modular multilevel converter based battery-ultracapacitor hybrid energy storage system for photovoltaic applications. In: 2015 Clemson university power systems conference, PSC 2015. Art. no. 7101674

61. Branco H, Castro R, Setas Lopes A (2018) Battery energy storage systems as a way to integrate renewable energy in small isolated power systems. Energy Sustain Dev 43:90–99. http://www.elsevier.com, https://doi.org/10.1016/j.esd.2018.01.003

62. Rastler D (2010) Electricity energy storage technology options: a white paper primer on applications, costs, and benefits. Electric Power Research Institute (EPRI), Technical update, USA

63. Divya KC, Østergaard J (2009) Battery energy storage technology for power systems—an overview. Electric Power Syst Res 79(4):511–520. https://doi.org/10.1016/j.epsr.2008.09.017

64. Ea Technology (2004) Review of electrical energy storage technologies and of their potential for the UK. Contract Number, URN Number 04/1876, pp 1–34

65. Kosin L, Usach F (1995) Electric characteristics of lead battery. Russ J Appl Chem 143(3):1–4

66. Chan HL, Sutanto D (2000) A new battery model for use with, battery energy storage systems and electric vehicles power systems. IEEE

67. Ceraolo M (2000) Dynamical models of lead-acid batteries. IEEE Trans Power Syst 15:1184–1190

Chapter 5
Design of Hybrid Renewable Energy Systems

5.1 Introduction

The design of a PV, wind or hybrid system can be made based on the exact knowledge of the load, the absorbed solar radiation, the estimates surface to be installed (especially for PV panels) and the choice of other equipment (controllers, inverters).

5.2 Design of Photovoltaic Systems

The effectiveness of any electric system depends on its design and its use. The sizing should be based on meteorological data, solar radiation and the exact load profile of consumers over long periods.

5.2.1 Determination of the Load Demand of Consumers

The exact knowledge of the customers load demand determines the size of generators [1].

$$D_{energy-total} = \sum P_{Load} . t \qquad (5.1)$$

where P_{load} is the load power and t is run time (hours) per day (Fig. 5.1).

© Springer Nature Switzerland AG 2020
D. Rekioua, *Hybrid Renewable Energy Systems*, Green Energy and Technology,
https://doi.org/10.1007/978-3-030-34021-6_5

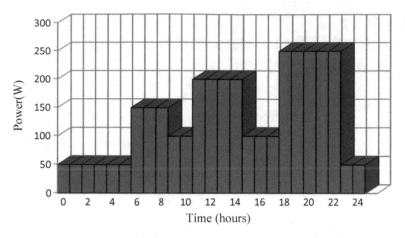

Fig. 5.1 Example of a time schedule diagram

Application 1

For example, in this case, the daily energy is calculated as:

$$D_{energy} = (50\,W) \times (5\,h) + (150\,W) \times (3\,h) + (100\,W) \times (2\,h)$$
$$+ (200\,W) \times (4\,h) + (100\,W) \times (3\,h) + (250\,W)$$
$$\times (5\,h) + (50\,W) \times (2\,h)$$
$$D_{energy} = 3350\,Wh/day$$

It can be also calculated by knowing the different appliances, their power and their run time. For example, (see Table 5.1). In this example, it is listed some appliances used in a house to supply them with renewable energy (solar, Wind,...).

Table 5.1 Results of Application 2

Appliances	Number (N)	Power (W) P	Run time (h/day)t	Daily energy (Wh/day) D_{energy}
Oven	1	500	1	500
Steam iron	1	850	1	850
Washing machine	1	300	1	300
Television	1	200	6	200
Laptop computer	1	30	6	180
Water pump	1	400	2	800
Lights	2	12	5	120
Hair dryer	1	400	1	300
			Total Daily energy (Wh/day)$D_{energy\text{-}total}$	**3050**

The total daily energy will be calculated as:

$$D_{\text{energy\_total}} = N \cdot P \cdot t \tag{5.2}$$

Application 2
For example, we want to supply a house with solar energy. The different electrical devices are listed in Table 5.1.

5.2.2 Photovoltaic System Design

Once the load and absorbed solar radiation are known, the design of the PV system can be carried out, including the estimation of the required PV panel's area and the selection of the other equipment (controllers, inverters,…).

Different methods have been used for designing PV systems. Each method depends on specifically output parameters.

5.2.2.1 First Method

This method is based on the load demand. The actual daily solar energy is given as:

$$P_{a-\text{PV}/\text{day}} = P_p \cdot \frac{E_{s-\text{Worst}}}{E_{s-\text{STC}}} \left(1 - \sum \text{losses} \right) \tag{5.3}$$

with: $P_{a-\text{PV}/\text{day}}$: actual daily power, P_p: peak power of panels, $E_{s-\text{Worst}}$: value of the monthly average irradiation of the worst month of irradiation, $E_{s-\text{STC}}$ the irradiation value under STC conditions (standard test conditions), $\sum \text{losses} = 20\%$.

Hence, the PV panel number is:

$$N_{\text{pv}} = \frac{D_{\text{energy}-\text{total}}}{P_{a-\text{PV}/\text{day}}} \tag{5.4}$$

The total peak power will be:

$$P_{p-\text{total}} = N_{\text{pv}} \cdot P_p \tag{5.5}$$

Ant the total PV panel area is:

$$A_{\text{pv}-\text{total}} = N_{\text{pv}} \cdot A_{\text{pv}-u} \tag{5.6}$$

with: $A_{\text{pv}-u}$ is the unit PV panel area (m^2), $A_{\text{pv}-\text{tot}}$ the estimated total PV area (m^2).

Application 3

$$D_{\text{energy\_total}} = 500 \, \text{Wh/day}, E_{s-\text{Worst}} = 2.73 \, \text{kWh/m}^2 \cdot \text{day}, E_{s-\text{STC}}$$
$$= 1000 \, \text{W/m}^2, A_{\text{pv}-u} = 1.4 \, \text{m}^2.$$

The obtained results for three different photovoltaic powers can be summarized in Table 5.2.

5.2.2.2 Second Method

This method is very simple but remains an estimated one. It is necessary to know the need energy, the PV efficiency (material) and the value of the radiation of the most unfavorable month of the site.

$$P_{\text{pv}-\text{totale}-\text{est}} = \frac{D_{\text{energy\_total}} \cdot E_{\text{STC}}}{E_{\text{worst}}} \tag{5.7}$$

where h_{sun} is the peak sun-hour can be written as:

$$h_{\text{sun}} = \frac{E_{\text{worst}}}{E_{\text{STC}}} \tag{5.8}$$

Thus:

$$P_{\text{pv}-\text{totale}-\text{est}} = \frac{D_{\text{energy\_total}}}{h_{\text{sun}}} \tag{5.9}$$

Application 4

$$E_{s-\text{Worst}} = 2.73 \, \text{kWh/m}^2.\text{day}, E_{s-\text{STC}} = 1000 \, \text{W/m}^2, A_{\text{pv}-u} = 1.4 \, \text{m}^2.$$

See (Table 5.3).

Table 5.2 Results of the Application 3

$D_{\text{energy\_total}}$ (Wh/day)	$P_p(\text{W}_p)$	$E_{s-\text{worst}}(\text{kWh/m}^2/\text{day})$	$P_{\text{a}-\text{PV}/\text{day}}(W)$	N_{pv}	$P_{\text{pv}-\text{total}}(W)$	$A_{\text{pv}-\text{total}}$
500	80	2.73	174.72	3	240	4.2
	100		218.4	3	300	4.2
	160		349.44	2	320	2.8

Table 5.3 Results of the Application 4

D_{energy_total} Wh/day	$P_p(W_p)$	h_{sun} (hour)	$P_{pv-totale-est}$	N_{pv}	$P_{pv-total}(W)$	$A_{pv-total}(m^2)$
500	80	2.73	183.1501832	3	240	4.2
	100		183.1501832	2	200	2.8
	160		109.8901099	1	160	1.4

5.2.2.3 Third Method

This method is based on the monthly average solar irradiance. The monthly energy produced by the system per unit area is denoted $E_{pv,m}$ (kWh/m^2) and $M_{energy,m}$ is the monthly energy required by the load (where m = 1, 2, ..., 12 represents the month of the year.). The minimum surface of the generator needed to ensure full (100%) energy (F_{energy}) is expressed by [1]:

$$A_{pv-tot} = \frac{M_{energy,m}}{E_{pv,m}} \tag{5.10}$$

The full energy can be given by:

$$F_{energy} = E_{pv} \cdot A_{pv-tot} \tag{5.11}$$

The number of photovoltaic generators is calculated using the surface of the system unit A_{pv-u} taking the entire value:

$$N_{pv} = \frac{A_{pv-tot}}{A_{pv-u}} \tag{5.12}$$

Application 5
An application is made with the different parameters: $A_{pv,\,u}$ = 1.4 m^2, η_{pv} = 0.12, P_{pv} = 80 W$_p$ (Table 5.4).

5.2.2.4 Method Based on Load Needs

The number of the series-connection PV modules is calculated by:

$$N_{pv-serial} = \frac{F_{energy}}{E_{worst} \cdot \eta_{batt} \cdot \eta_{el} \cdot \eta_{DC}} \tag{5.13}$$

η_{batt} is the efficiency of the battery, η_{el} is the electrical efficiency of the whole installation (charge controller, inverter...), η_{DC} is the distribution circuit.

Table 5.4 Results of the application 5

Months	G (kWh/m^2/day)	$M_{energy,m}$ (Wh)	F_{energy} (Wh/m^2/day)	$A_{pv,tot,m}$ (m^2)	N_{pv}
January	2.38	12,395.04	37,200.00	3.00	3.00
February	3.31	16,126.32	34,800.00	2.16	2.00
March	4.44	23,123.52	37,200.00	1.61	2.00
April	5.46	27,518.40	36,000.00	1.31	1.00
May	6.41	33,383.28	37,200.00	1.11	1.00
June	7.12	35,884.80	36,000.00	1.00	1.00
July	7.23	37,653.84	37,200.00	0.99	1.00
August	6.38	33,227.04	37,200.00	1.12	1.00
September	5.08	25,603.20	36,000.00	1.41	2.00
October	3.66	19,061.28	37,200.00	1.95	2.00
November	2.51	12,650.40	36,000.00	2.85	3.00
December	2.06	10,728.48	37,200.00	3.47	3.00

It can be also written as:

$$N_{pv-serial} = \frac{F_{energy}}{E_{worst} \cdot K_E} \tag{5.14}$$

where $K_E = \eta_{batt} \cdot \eta_{el} \cdot \eta_{DC}$ is energy efficiency, it varies [0.6–0.75].

The maximum terminal voltage of the photovoltaic generator is estimated by:

$$V_{pv-max} = 1.15 \cdot N_{pv-serial} \cdot V_{oc} \tag{5.15}$$

(1.15 is a correction factor).

The PV parallel panels can be calculated by using Eq. 5.15, where V_{DC-bus} is DC bus voltage

$$N_{pv-para} = \frac{U_{pv-max}}{V_{DC-bus}} \tag{5.16}$$

Then, the total number of panels is deduced:

$$N_{pv} = N_{pv-para} \cdot N_{pv-serial} \tag{5.17}$$

The total photovoltaic power to be installed will be:

$$P_{pv-totale} = N_{pv} \cdot P_p \tag{5.18}$$

Application 6

The obtained results for three different photovoltaic powers can be summarized in Table 5.5.

Table 5.5 Results of the Application 6

F_{energy} (Wh/day)	P_p (W_p)	V_{oc} (V)	$N_{pv-serial}$	V_{pv-max} (V)	$N_{pv-para}$	N_{pv}
500	80	22.4	1	25.76	3	3
	100	20.5	1	23.575	2	2
	160	21.8	1	25.07	3	3

5.3 Design of Wind System

5.3.1 Calculation of Wind Energy

The energy produced by the wind generator during a period time Δt is expressed by:

$$E_{wind} = P_{mec} \cdot \Delta t \tag{5.19}$$

where Δt period of time.

5.3.2 Determination of the Wind Generator Size

The total area of the wind turbine generators required to ensure full coverage (100%) of the load (F_{energy}) is expressed by:

$$A_{wind-total} = \frac{D_{energy}}{E_{wind}} \tag{5.20}$$

with: $E_{wind}(kWh/m^2)$ is the monthly energy produced by the wind system per unit area and $M_{energy}(kWh)$ represents the monthly energy required by the load.

The number of wind turbine generators is calculated according to the surface area of the system unit by taking the entire value of the excess ratio.

$$N_{wind} = \frac{A_{wind-total}}{A_{wind}} \tag{5.21}$$

with A_{wind} is the surface area of a wind turbine.

5.4 Sizing of Hybrid Photovoltaic/Wind System

The energy produced by a photovoltaic generator per unit area is estimated using data from the global irradiance on an inclined plane, ambient temperature and the data sheet for the used photovoltaic pannels. It is given by:

$$E_{pv} = \eta_{pv} \cdot GA_{pv} \tag{5.22}$$

where: G is the solar radiation incident.

The power contained in the form of kinetic energy per unit area in the wind is expressed by:

$$P_{mec} = \frac{1}{2} \cdot \rho_{air} \cdot v_{wind}^3 C_p \cdot A_{wind} \tag{5.23}$$

with: v_{wind} is the speed wind, C_p the power coefficient, ρ_{air} the air density and A_{wind} the wind area.

- Pre-sizing of photovoltaic and wind systems:

The monthly energy produced by the system per unit of area is denoted $E_{pv,m}$ (kWh/m^2) for photovoltaic energy and $E_{wind,m}$ (kWh/m^2) for wind energy and $E_{L,m}$ M_{energy} represents the energy required by load every month (where m=1, 2, …, 12 represents the month of the year). Owe have:

$$E_{pv,m} = \sum_{month\ m} \Delta E_{pv} \tag{5.24}$$

$$E_{wind,m} = \sum_{month\ m} \Delta E_{wind} \tag{5.25}$$

and

$$F_{energy} = \sum_{month\ m} M_{energy} \tag{5.26}$$

Pre-sizing is sometimes based on the worst month of the year.

$$E_{L,worst\ m} = E_{pv,worst\ m} \cdot A_{pv} + E_{wind,worst\ m} \cdot A_{wind} \tag{5.27}$$

The parameter f which is the fraction of load supplied by the photovoltaic energy is introduced, $(1 - f)$ being the fraction of load supplied by the wind energy. Then:

$f = 1$ indicates that the entire load is supplied by the photovoltaic source.
$f = 0$ indicates that the entire load is powered by the wind source.

The different PV and wind area can be calculated as:

$$A_{pv} = \frac{fE_{L,worst\ m}}{E_{pv,worst\ m}} \tag{5.28}$$

$$A_{wind} = \frac{(1-f)E_{L,worst\ m}}{E_{wind,worst\ m}} \tag{5.29}$$

The pre-sizing is often also based on monthly annual average [2, 3]. The calculation of the size of wind generator and photovoltaic (A_{pv} and A_{wind}) is established from the annual average values of each monthly contribution $\left(\overline{E_{pv}} \text{ and } \overline{E_{wind}}\right)$. The load is represented by the monthly annual average $\overline{F_{energy}}$.

$$A_{pv} = f \cdot \frac{\overline{F_{energy}}}{\overline{E_{pv}}}$$
$$A_{wind} = (1-f) \cdot \frac{\overline{F_{energy}}}{\overline{E_{wind}}}$$

(5.30)

The number of photovoltaic and wind generator to consider is calculated according to the area of the system unit taking the integer value of the ratio by excess.

$$N_{pv} = \text{ENT}\left[\frac{A_{pv}}{A_{pv,u}}\right]$$
$$N_{wind} = \text{ENT}\left[\frac{A_{wind}}{A_{wind,u}}\right]$$

(5.31)

Application 7

Table 5.6 shows the monthly energy production of the generators and the size required to satisfy a constant daily consumption load of about 3050 Wh/day.

Table 5.6 Monthly energies produced by photovoltaic and wind generators

Months	G (kWh/ m²/day)	V_{wind} (m/s)	$M_{energy-pv,m}$ (kWh/day)	$M_{energy-wind,m}$ (kWh/day)	F_{energy} (kWh/m²/ day)	A_{pv} (m²)	A_{wind} (m²)
January	2.38	5.22	11.56	14.97	153.45	6.70	4.87
February	3.31	5.32	11.80	13.90	143.55	6.14	4.91
March	4.44	5.3	13.51	14.32	153.45	5.73	5.09
April	5.46	5.34	13.34	12.75	148.5	5.62	5.53
May	6.41	4.52	13.68	11.35	153.45	5.66	6.42
June	7.12	4.33	13.71	12.27	148.5	5.46	5.75
July	7.23	4.46	14.48	13.28	153.45	5.35	5.49
August	6.38	4.36	14.65	13.89	153.45	5.28	5.25
September	5.08	4.17	13.91	11.25	148.5	5.39	6.27
October	3.66	4.48	13.13	12.29	153.45	5.90	5.93
November	2.51	5.15	11.31	12.95	148.5	6.63	5.45
December	2.06	5.33	10.91	17.88	153.45	7.10	4.08
Monthly average	4.67	4.83	13.00	13.42	150.975		

Table 5.7 Sizing according to the annual monthly average

f	$1 - f$	A_{pv} (m^2)	N_{pv}	A_{wind} (m^2)	N_{wind}	F_{moy}
0	1.00	0.00	0	9.42	3	126.4164
0.1	0.90	1.29	2	9.42	3	143.1864
0.2	0.80	1.94	3	9.42	3	151.6364
0.3	0.70	3.23	5	9.42	3	168.4064
0.4	0.60	3.88	6	6.28	2	134.7176
0.5	0.50	5.17	8	6.28	2	151.4876
0.6	0.40	5.81	9	6.28	2	159.8076
0.7	0.30	7.11	11	3.14	1	134.5688
0.8	0.20	7.75	12	3.14	1	142.8888
0.9	0.10	8.40	13	3.14	1	151.3388
1	0.00	9.69	15	0.00	0	

After having calculated the total required surfaces of the two generators (photovoltaic, wind), we will determine the number to install according to the fraction of the load (f) taken with an interval of [0–1] (Table 5.7).

The obtained results can be represented in Fig. 5.2.

5.5 Sizing of Hybrid Photovoltaic/Wind System/Batteries

The global system includes a PG generator, a charge controller, a battery bank and a DC/AC converter supplying a load profile. These major ¡components should be selected to the location site and the application. Figure 5.3 shows a diagram of a typical standalone PV system powering AC loads [5–32].

5.5.1 Battery Design

Battery capacity is the energy per day capable of charging a battery. The calculation can be written as:

$$C_{batt}(A \cdot h) = \frac{D_{energy} \cdot N_{aut}}{V_{batt} \cdot DOD \cdot \eta_{batt}} \tag{5.32}$$

where: V_{batt} is the battery voltage, DOD is the depth of discharge, η_{batt} the efficiency battery, N_{aut} is the days of autonomy and D_{energy} is the total energy required.

The number of batteries to be used is determined from the capacity of a battery unit $C_{batt,u}$ is given by:

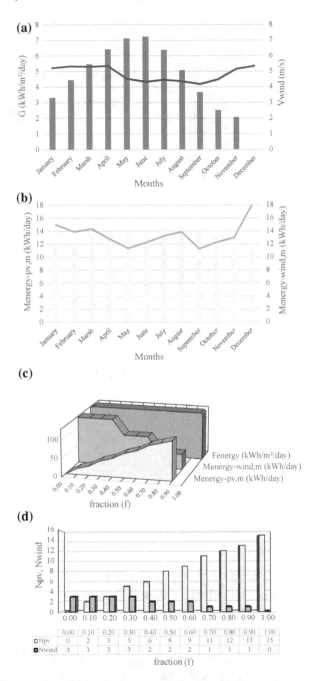

Fig. 5.2 Obtained results of PV/wind design. **a** Solar radiation and wind speeds of the location. **b** Average daily energy of PV and wind generators. **c** Full energy with PV and wind turbine energies. **d** Combinations of number of PV and wind turbines

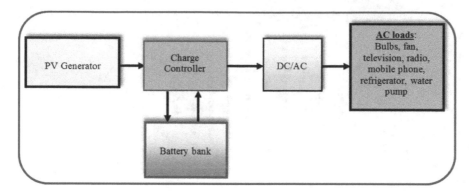

Fig. 5.3 Diagram of a typical standalone PV system powering AC loads

Table 5.8 Results of Application 9

D_{energy} (Wh/day)	V_{batt} (V)	η_{batt}	DOD	N_{aut} (h/day)	C_{batt} (Ah)	$C_{batt,u}$ (Ah)	N_{batt}
300	12	0.85	0.6	2	98.04	90	1
3030					990.1		11

$$N_{batt} = \left\lceil \frac{C_{batt,min}}{C_{batt,u}} \right\rceil \tag{5.33}$$

Application 8

We make an application for two different daily energies (300 Wh/day and 3030 Wh/day). (Table 5.8).

5.5.2   DC/AC Converter Design

5.6   Design of Hybrid Photovoltaic/Wind System/Fuel Cells

In this case, PV and wind turbine generators are considered as a main source and fuel cells as a secondary one.

$$P_{Ren} = P_{PV} + P_{wind} \tag{5.34}$$

$$P_{\text{Load}} = P_{\text{Ren}} + P_{\text{FC}} \tag{5.35}$$

where: P_{Ren} is the power produced by PV and wind systems and P_{FC} is the power produced by fuel cell system.

The design of PV panels and wind turbine has been explained in Sect. 5.4. For fuel cells, the sizing methodology is as follows: Stack design consists of calculating the number and area of cells that make up a fuel cell stack. This sizing must take into account the nominal power of the cell and the current density desired to obtain by adding 20% of the power that will be consumed by the cell auxiliaries.

5.6.1 Power Calculation

$$P_{\text{FC-stack}} = P_{\text{inv}} \cdot (1 + 0.2) \tag{5.36}$$

where P_{inv} *is* the input inverter power.

5.6.2 Cell Number and Cell Surface

The maximum electrical power of the stack is calculated by the following relationship [4]:

$$P_{\text{FC-stack}} = N_{\text{FC}} \cdot E_{\text{FC}} \cdot j \cdot A_{\text{FC}} \tag{5.37}$$

with: $P_{\text{FC-stack}}$ is the maximum electrical power of the stack (W), N_{FC} is the number of cells in the stack, E_{FC} is the voltage per cell (V), j is the current density (A/m^2) and A_{FC} is the active cell area (m^2).

The voltage of the stack depends on the cells number:

$$U_{\text{stack}} = N_{\text{FC}} \cdot E_{\text{FC}} \tag{5.38}$$

To determine the area of the stack, the FC current must first be calculated:

$$I_{\text{stack}} = \frac{P_{\text{stack}}}{U_{\text{stack}}} \tag{5.39}$$

$$A_{\text{FC}} = I_{\text{stack}}/j \tag{5.40}$$

Table 5.9 Results of application 8

$V_{DC-bus}(V)$	$E_{FC}(V)$	$j(A/cm^2)$	$U_{stack}(V)$	N_{FC}	$P_{inv}(W)$	$P_{FC-stack}(W)$	$I_{stack}(A)$	$A_{FC}(cm^2)$
450	1	1	225	376	1500	1800	8	13.33

It is interesting to have the highest voltage U_{stach} and, therefore, the lowest current I_{stach} because it limits losses in the cell.

Application 9

An example is made to follow the sizing method (Table 5.9).

5.7 Application to Water Pumping System

Depending on the size of pump, three-phase induction machines or single-phase induction machines can be used. In this work, induction motor associated with a centrifugal pump has been used. The energy consumed by the pump depends on the desired water flow, which represents the energy that must be provided by the two generators (PV and wind).

Sizing of the different components of the system supplying a small village with water has been made. The specifications must satisfy the following conditions:

- the volume of water tank pumped per day about 100 m^3.
- the water tank is situated at 10 m above the surface level.
- a nominal flow rate of 34 m^3/h=0.0094 m^3/s.

The different results can be summarized in Table 5.10.

As the height increases, the powers increase, which will improve efficiency (Fig. 5.4).

Table 5.10 Moto-pump group sizing

Symbols	Equations	Results
Hydraulic power P_{Hyd}	$P_{Hyd} = \rho_{water} \cdot g \cdot h \cdot q_v$	922.14 W
Mechanical power required by the pump P_{mec}	$P_{mec} = \frac{P_{Hyd}}{\eta_{pump}}$	2049.20 W
Electrical power required for the motor to operate P_{elec}	$P_{elec} = \frac{P_{mec}}{\eta_{motor}}$	2561.50 W
Input inverter power P_{inv}	$P_{inv} = \frac{P_{elec}}{\eta_{inv}}$	2696.32 W
Pumping time required to satisfy the water needs τ_{pump}	$\tau_{pump} = \frac{V_{tank}}{q_v}$	2.94 h
Daily electrical energy required E_c	$D_{energy} = \tau_{pump} \cdot P_{inv}$	7930.34 Wh/day

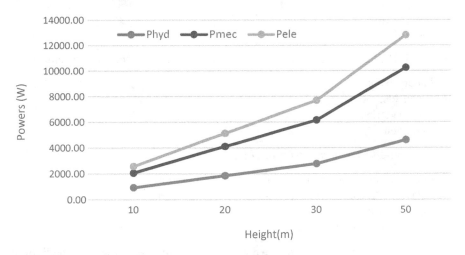

Fig. 5.4 Variation of powers

5.8 Optimization of Power System Using HOMER Pro Software

5.8.1 Introduction to HOMER Pro Software

Hybrid optimization model for electric renewable (HOMER) software performs economic analysis on hybrid power systems. Homer is a simulation and optimization software for multi-source (hybrid) power generation systems, with different components: (PV, wind, grid, storage, diesel…). It is dedicated directly to the simulation of on-grid and off-grid systems. The software allows the simulation of a system based on inputs (solar, wind, diesel, etc.) according to energy consumption. Subsequently, it is possible to analyze several different configurations for the same system in order to obtain a cost-effective system. The software simulates all the required configurations and gives the best solution, the cheapest solution, among them. Then, it is finally possible to perform sensitivity analyses to determine if the solution found is the best even if there are some changes in the various parameters entered (variation in the cost of the technology, variation in the deposit data, etc.). It is, therefore, possible to perform many analyses with many different configurations in less than a few minutes of simulation.

The software allows simulations to be performed with different energy production systems:

- photovoltaic solar panels,
- wind turbines,
- hydro power,
- biomass,
- generators (diesel, gasoline, biogas, alternative and customized fuels),

– power grid,
– fuel cells.

HOMER also offers a wide range of energy storage or recovery systems:

– battery bank,
– flywheels,
– flow batteries,
– hydrogen.

You can also input various types of energy needs:

– daily consumption profiles with seasonal variations,
– delayed charging for water pumping or refrigeration,
– thermal load,
– energy efficiency measures.

HOMER can, therefore, simulate a wide range of different systems in addition to all possible combinations of hybrid systems (Fig. 5.5).

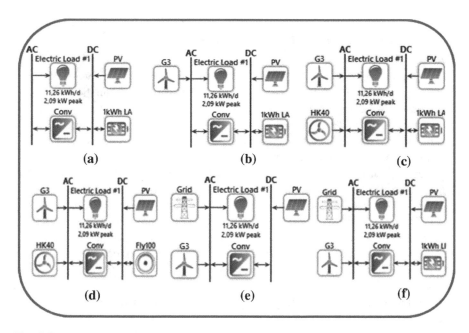

Fig. 5.5 Example hybrid combinations systems in Homer Pro. **a** PV/batteries. **b** PV/wind/batteries. **c** PV/wind/batteries/HPS. **d** PV/wind/HPS/FEES. **e** PV/wind/batteries on grid. **f** PV/wind/batteries/batteries on grid

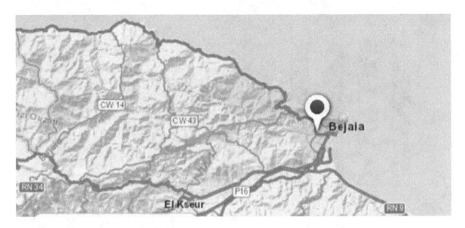

Fig. 5.6 Bejaia location in Algeria (Latitude 36°45.3522′ N, Longitude 5°5.0598′ E) with HOMER software License Agreement

5.8.2 Application to PV/Wind System with Battery Storage

The main purpose of this application is to optimize the size of PV/wind/wind/ battery hybrid system components, minimize excess production and perform a cost analysis based on life-cycle cost. Solar radiation and wind speed data were collected for Bejaia area in Algeria (Latitude 36°45.3522′ N, Longitude 5°5.0598′ E) using

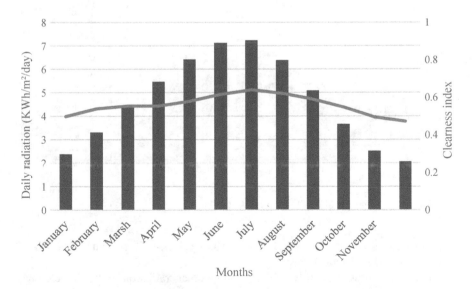

Fig. 5.7 Daily radiation and clearness index at Bejaia location (downloaded at 18/08/2019 18:58:26 from HOMER software License Agreement)

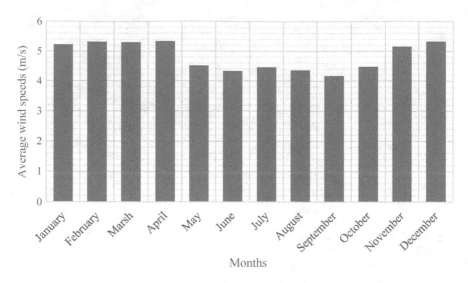

Fig. 5.8 Average wind speeds (downloaded at 18/08/2019 18:58:26 from HOMER software License Agreement)

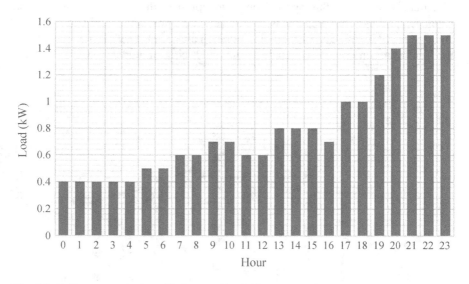

Fig. 5.9 Daily average load profile for a residential house

the function of "Get Data via Internet" in the HOMER software (the NASA Atmospheric Data Center), (see Fig. 5.6).

Figure 5.7 shows monthly average daily solar radiation with the clearness radiation at Bejaia site.

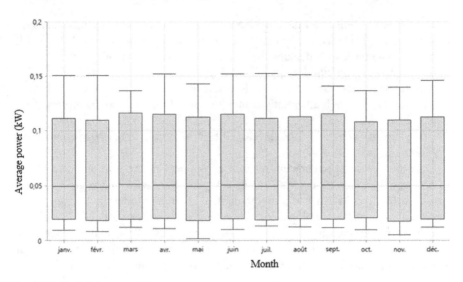

Fig. 5.10 Scaled data monthly average

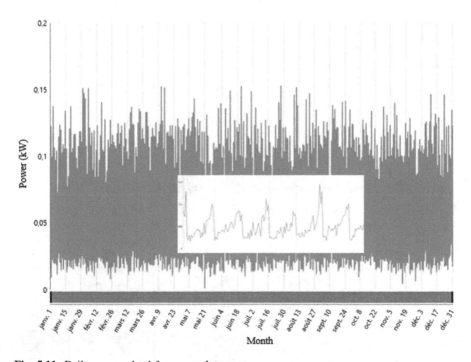

Fig. 5.11 Daily average load for a complete year

The average wind speeds are as follows in Fig. 5.8. It varies from 4.170 to 5.33 (m/s). The annual average is about 4.83 m/s.

In this application, load data were collected for a residential house located in Bejaia region. The measured annual consumption is estimated as 1.2 kWh/day (Fig. 5.9). The peak load decides the size of system components.

The scaled data monthly average is as Fig. 5.10.

The daily average load for a complete year is (Fig. 5.11).

The hourly average load variations in a year for all months can be represented as (Fig. 5.12).

The main components of the developed hybrid system under Homer are shown in Fig. 5.13.

Once the technical parameters of each component are chosen, the cost of each component is entered by entering the initial price, the maintenance price and their estimated lifetime, in order to allow the software to determine the overall price of the installation and to optimize for the lowest net present cost (NPC). HOMER calculates the net present cost of each component and of the hybrid system as a whole. The results were computed in different simulations to show the technico-economic feasibility of the studied hybrid system.

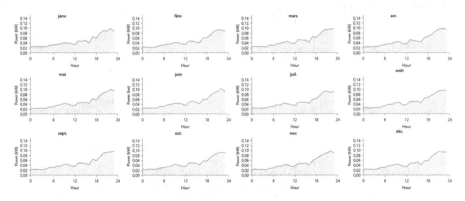

Fig. 5.12 Hourly average load variations in a year for all months

Fig. 5.13 Block diagram of
PV/wind turbine/battery
hybrid system

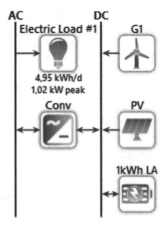

5.9 Conclusions

The most presented methods are well-known but still used in the design of system because it is the most important step in a project. This application using Homer software is based on economical performances which depend of course on real market prices and exact sizing of each component of a studied system.

References

1. Deshmukh MK, Deshmukh SS (2008) Modeling of hybrid renewable energy systems. Renew Sustain Energy Rev 12(1):235–249. https://doi.org/10.1016/j.rser.2006.07.011
2. Luu T, Nasiri A (2010) Power smoothing of doubly fed induction generator for wind turbine using ultracapacitors. In: IECON proceedings (industrial electronics conference), art no 5675040, pp 3293–3298
3. An Y, Fang W, Ming B, Huang Q (2015) Theories and methodology of complementary hydro/photovoltaic operation: applications to short-term scheduling. J Renew Sustain Energy 7(6):063133. https://doi.org/10.1063/1.4939056. http://scitation.aip.org/content/aip/journal/jrse
4. Qandil MD, Abbas AI, Qandil HD, Al-Haddad MR, Amano RS (2019) A stand-alone hybrid photovoltaic, fuel cell, and battery system: case studies in Jordan. J Energy Resour Technol, Trans ASME 141(11):111201
5. Ardakani FJ, Riahy G, Abedi M (2010) Design of an optimum hybrid renewable energy system considering reliability indices. In: Proceedings—2010 18th Iranian Conference on Electrical Engineering, ICEE 2010, art no 5506958, pp 842–847. https://doi.org/10.1109/iraniancee.2010.5506958
6. Belmili H, Haddadi M, Bacha S, Almi MF, Bendib B (2014) Sizing stand-alone photovoltaic-wind hybrid system: techno-economic analysis and optimization. Renew Sustain Energy Rev 30:821–832. https://doi.org/10.1016/j.rser.2013.11.011
7. Nogueira CEC, Vidotto ML, Niedzialkoski RK, De Souza SNM, Chaves LI, Edwiges T, Santos DBD, Werncke I (2014) Sizing and simulation of a photovoltaic-wind energy system using batteries, applied for a small rural property located in the south of Brazil. Renew Sustain Energy Rev 29:151–157. https://doi.org/10.1016/j.rser.2013.08.071
8. Feroldi D, Zumoffen D (2014) Sizing methodology for hybrid systems based on multiple renewable power sources integrated to the energy management strategy. Int J Hydrogen Energy 39(16):8609–8620. https://doi.org/10.1016/j.ijhydene.2014.01.003
9. Iverson Z, Achuthan A, Marzocca P, Aidun D (2013) Optimal design of hybrid renewable energy systems (HRES) using hydrogen storage technology for data center applications. Renew Energy 52:79–87. https://doi.org/10.1016/j.renene.2012.10.038
10. Jallouli R, Krichen L (2012) Sizing, techno-economic and generation management analysis of a stand alone photovoltaic power unit including storage devices. Energy 40(1):196–209. https://doi.org/10.1016/j.energy.2012.02.004. www.elsevier.com/inca/publications/store/4/8/3/
11. Kashefi Kaviani A, Riahy GH, Kouhsari SHM (2009) Optimal design of a reliable hydrogen-based stand-alone wind/PV generating system, considering component outages. Renew Energy 34(11):2380–2390. https://doi.org/10.1016/j.renene.2009.03.020
12. Luna-Rubio R, Trejo-Perea M, Vargas-Vázquez D, Ríos-Moreno GJ (2012) Optimal sizing of renewable hybrids energy systems: a review of methodologies. Solar Energy 86(4):1077–1088. https://doi.org/10.1016/j.solener.2011.10.016

13. Erdinc O, Uzunoglu M (2012) Optimum design of hybrid renewable energy systems: overview of different approaches. Renew Sustain Energy Rev 16(3):1412–1425. https://doi.org/10.1016/j.rser.2011.11.011
14. Yang H, Wei Z, Chengzhi L (2009) Optimal design and techno-economic analysis of a hybrid solar-wind power generation system. Appl Energy 86(2):163–169. https://doi.org/10.1016/j.apenergy.2008.03.008.   http://www.elsevier.com/inca/publications/store/4/0/5/8/9/1/index.httdoi
15. Diaf S, Diaf D, Belhamel M, Haddadi M, Louche A (2007) A methodology for optimal sizing of autonomous hybrid PV/wind system. Energy Policy 35(11):5708–5718. https://doi.org/10.1016/j.enpol.2007.06.020
16. Abdilahi AM, Mohd Yatim AH, Mustafa MW, Khalaf OT, Shumran AF, Mohamed Nor F (2014) Feasibility study of renewable energy-based microgrid system in Somaliland's urban centers. Renew Sustain Energy Rev 40:1048–1059. https://doi.org/10.1016/j.rser.2014.07.150
17. Hosseinalizadeh R, Shakouri GH, Amalnick MS, Taghipour P (2016) Economic sizing of a hybrid (PV-WT-FC) renewable energy system (HRES) for stand-alone usages by an optimization-simulation model: case study of Iran. Renew Sustain Energy Rev 54:139–150
18. Mandal S, Das BK, Hoque N (2018) Optimum sizing of a stand-alone hybrid energy system for rural electrification in Bangladesh. J Clean Prod 200:12–27
19. Markvart T (1996) Sizing of hybrid photovoltaic-wind energy systems. Solar Energy 57 (4):277–281. https://doi.org/10.1016/s0038-092x(96)00106-5
20. Sba KM, Bakell Y, Kaabeche A, Khenfous S (2019) Sizing of a hybrid (photovoltaic/wind) pumping system based on metaheuristic optimization methods. In: 2018 International Conference on Wind Energy and Applications in Algeria, ICWEAA 2018, 8605053
21. Muhsen DH, Nabil M, Haider HT, Khatib T (2019) A novel method for sizing of standalone photovoltaic system using multi-objective differential evolution algorithm and hybrid multi-criteria decision making methods. Energy 1158–1175
22. Hadidian Moghaddam MJ, Kalam A, Nowdeh SA, Ahmadi A, Babanezhad M, Saha S (2019) Optimal sizing and energy management of stand-alone hybrid photovoltaic/wind system based on hydrogen storage considering LOEE and LOLE reliability indices using flower pollination algorithm. Renew Energy 1412–1434
23. Khiareddine A, Ben Salah C, Rekioua D, Mimouni MF (2018) Sizing methodology for hybrid photovoltaic/wind/hydrogen/battery integrated to energy management strategy for pumping system. Energy 153:743–762
24. Ould-Amrouche S, Rekioua D, Hamidat A (2010) Modelling photovoltaic water pumping systems and evaluation of their CO_2 emissions mitigation potential. Appl Energy 87 (11):3451–3459
25. Nordin ND, Rahman HA (2019) Comparison of optimum design, sizing, and economic analysis of standalone photovoltaic/battery without and with hydrogen production systems. Renew Energy 107–123
26. Liu J, Chen X, Cao S, Yang H (2019) Overview on hybrid solar photovoltaic-electrical energy storage technologies for power supply to buildings. Energy Convers Manage 103–121
27. Farahmand MZ, Nazari ME, Shamlou S (2017) Optimal sizing of an autonomous hybrid PV-wind system considering battery and diesel generator. In: 2017 25th Iranian Conference on Electrical Engineering, ICEE 2017, art no 7985194, pp 1048–1053
28. Dong W, Li Y, Xiang J (2016) Sizing of a stand-alone photovoltaic/wind energy system with hydrogen and battery storage based on improved ant colony algorithm. In: Proceedings of the 28th Chinese Control and Decision Conference, CCDC 2016, art no 7531788, pp 4461–4466
29. Elma O, Selamogullari US (2012) A comparative sizing analysis of a renewable energy supplied stand-alone house considering both demand-side and source side dynamics. Appl Energy 96:400–408
30. Rizzo R, Piegari L, Tricoli P, Munteanu C, Topa V (2009) Sizing of photovoltaic sources and storage devices for stand-alone power plants. In: 2009 IEEE Bucharest PowerTech: innovative ideas toward the electrical grid of the future, art no 5281869. https://doi.org/10.1109/ptc.2009.5281869

31. Upadhyay S, Sharma MP (2014) A review on configurations, control and sizing methodologies of hybrid energy systems. Renew Sustain Energy Rev 38:47–63. https://doi.org/10.1016/j.rser.2014.05.057

32. Jallouli R, Krichen L (2012) Sizing, techno-economic and generation management analysis of a stand-alone photovoltaic power unit including storage devices. Energy 40(1):196–209

Chapter 6
Power Management and Supervision of Hybrid Renewable Energy Systems

6.1 Introduction

A remarkable number of studies in the literature are dedicated to power management of hybrid power systems [1–67], including several alternative techniques such as intelligent approaches (ANN, FLC, etc.), optimal control approaches and model-based predictive controllers.

The purpose of the power management control (PMC) is to coordinate the different sources of a hybrid system, particularly their power exchange, in order to make the generated power controllable, to control the DC bus voltage, to ensure the continuity of the load supply and to decrease the cost of energy production. The power management strategy depends on the type of energy system and its components (Fig. 6.1).

6.2 Different Combinations of Hybrid Systems

The most used combinations of hybrid system were presented in Chap. 1 (Fig. 1.5). Mathematically, it can have 2^n combinations of hybrid systems [9].

6.3 Photovoltaic/Battery System

The study sizing has provided the number of PV panels and batteries used in this configuration (see Chap. 5) (Fig. 6.2).

© Springer Nature Switzerland AG 2020
D. Rekioua, *Hybrid Renewable Energy Systems*, Green Energy and Technology,
https://doi.org/10.1007/978-3-030-34021-6_6

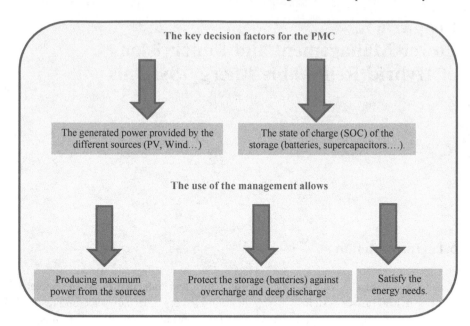

Fig. 6.1 Key decision factors for PMC

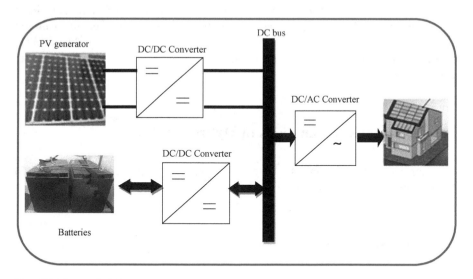

Fig. 6.2 Photovoltaic/battery system

6.3.1 Power Management Algorithm for the PV/Battery Installation (First Structure)

The key decision factor for this PMC is the power level provided by the PV generator and the state of charge of the batteries (SOC). The use of the management control allows a maximum power from the PV generator and protects the batteries against overcharge and depth discharge while satisfying the energy needs [1–5]. The power management shown in Fig. 6.3 depends on the three switches state [13, 31, 32] (K_1, K_2 and K_3).

To supervise the PV/battery installation, it is assumed that the batteries are initially charged. The management system mainly deals with the power supply of the load and the protection of the batteries (a minimum state of charge SOC_{min} and a maximum state of charge SOC_{max} are assumed). The different modes that control the operation of the proposed system are as follows (Fig. 6.4):

- **Mode 1**: The batteries are charged. The photovoltaic generator is sufficient to satisfy the load.
- **Mode 2**: The PV generator is insufficient to supply the load. The battery adds its power to satisfy the load.
- **Mode 3**: Only the battery supplies the load.
- **Mode 4**: The PV generator is sufficient to supply the load.
- **Mode 5**: Batteries completely discharged no photovoltaic production.

The power flow is represented as in Fig. 6.5.

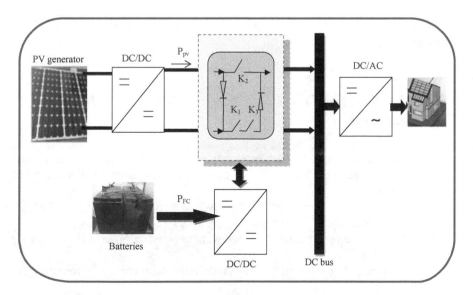

Fig. 6.3 Photovoltaic/battery system with PMC (first structure)

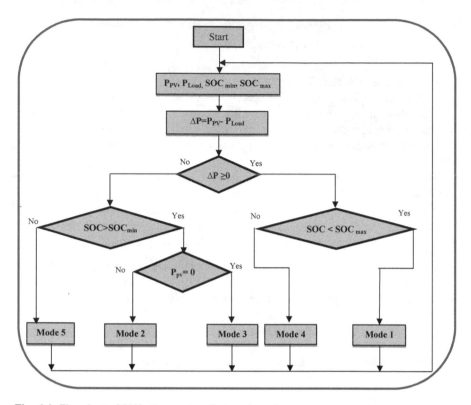

Fig. 6.4 Flowchart of PV/battery system-first configuration

Table 6.1 allows analyzing the different modes that can be distinguished during this operation.

6.3.2 Application of the First Structure Under MATLAB/ Simulink

To supervise this structure (PV/batteries system), a precise sizing has been made (see Chap. 5) to obtain the PV panels number, and battery number which are based on a daily energy (D_{energy}).

Example: N_{pv} = 20 panels of 80 Wp, N_{batt} = 6 batteries of 48 V, 200 Ah, D_{energy} = 2808 Wh/day.

At first, it is assumed that the batteries were charged at 50% of their capacity (SOC_{max} = 50%).

A chosen profile is given which corresponds to the daily energy for four different days (Fig. 6.6).

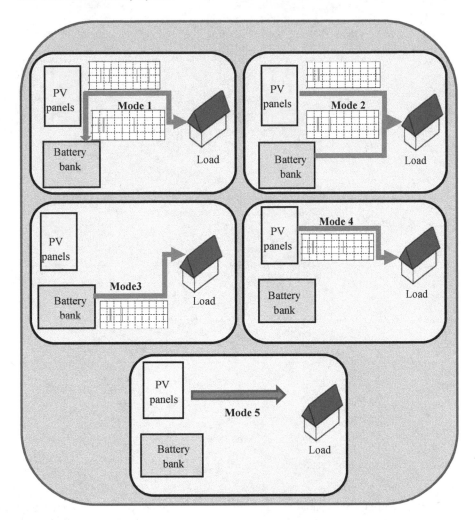

Fig. 6.5 Power flow of PV/batteries system (first configuration)

Table 6.1 Status of the switches and the different modes

Modes	K_1	K_2	K_3	P_{load}
Mode 1	On	On	Off	P_{pv}
Mode 2	On	Off	On	$P_{pv} + P_{batt}$
Mode 3	Off	Off	On	P_{bat}
Mode 4	On	Off	Off	P_{pv}
Mode 5	Off	Off	Off	Disconnection

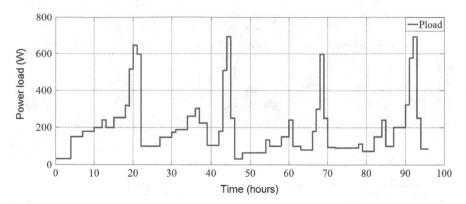

Fig. 6.6 Chosen load profile

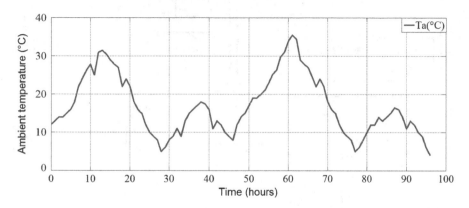

Fig. 6.7 Temperature profile during four different days

The measured solar radiation and temperature values of two different days (sunny and cloudy one) which were considered in our work are as follows (Figs. 6.7 and 6.8). The evolution of the different powers is shown in Figs. 6.9 and 6.10.

Conclusion: It has been noticed that there was a power excess in the first structure with only three switches used in the power supervision. It is therefore of interest to store this power in a dump load or an auxiliary load for another use or application (electrolyzer, moto-pump, etc.).

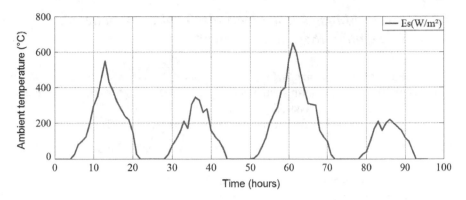

Fig. 6.8 Solar irradiance profile during four different days

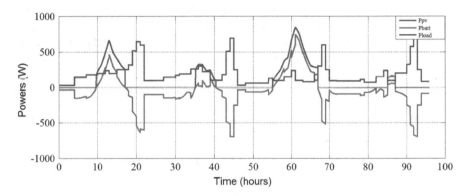

Fig. 6.9 Evolution of the different powers during four different days

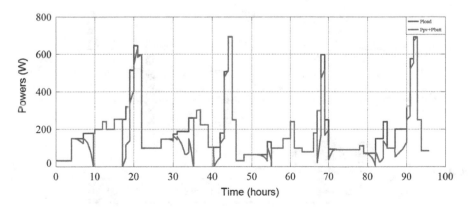

Fig. 6.10 Comparison between the sum of the different source powers and the load power

6.3.3 Power Management Algorithm for the PV/Battery Installation (Second Structure)

In this structure, four switches (K_1, K_2, K_3 and K_4) have been used for power supervision [9, 63]. The fourth switch is used to store the power excess if there is in a derivate or an auxiliary load (Fig. 6.11).

The system operates in one of the different modes (Fig. 6.12).

Mode 1: The photovoltaic power is sufficient to supply the load and charge batteries.

Mode 2: The photovoltaic power is insufficient; the power of batteries is added to satisfy the power load.

Mode 3: The batteries supply the load when no power provides from the PV generator.

Mode 4: The photovoltaic power is sufficient, and batteries are completely charging, so their disconnection is necessary to protect them.

Mode 5: The produced photovoltaic power is sufficient to supply the load and batteries are fully charged, so the excess energy will be dissipated in a derivate load or in an auxiliary source.

Mode 6: The produced photovoltaic power is insufficient to supply the load, and the batteries are discharged.

Mode 7: In this case, there is no PV power production (during night or cloudy day), and batteries are completely discharged, so the load will be disconnected.

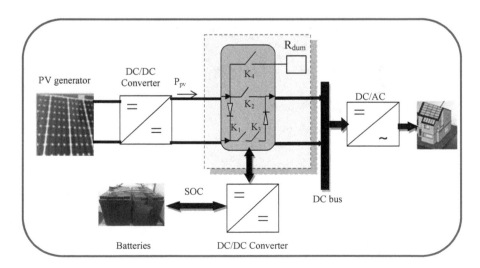

Fig. 6.11 Power flow of PV/battery system (second structure)

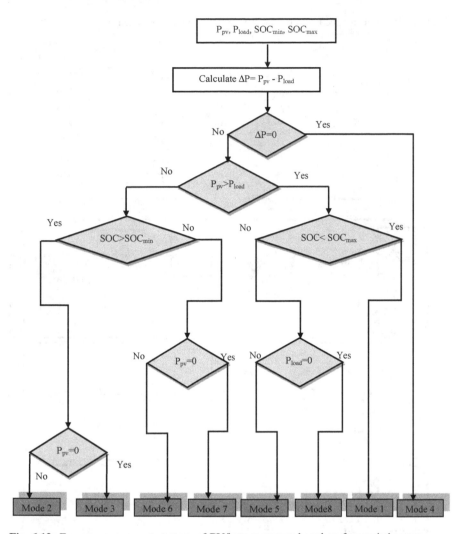

Fig. 6.12 Energy management strategy of PV/battery system based on four switches

Mode 8: The produced photovoltaic power is sufficient to supply the load and batteries are fully charged, but the load is disconnected, so the excess energy will be dissipated in a derivate or dump load.

Table 6.2 allows analyzing the different modes that can be distinguished during this operation.

The power flow is represented as in Fig. 6.13 for the second structure.

Table 6.2 Simplified table based on four switches of PV/battery system

K_1	K_2	K_3	P_{pv}	P_{batt}	P_{load}	Modes
0	0	0	0	0	0	Mode 1
0	0	0	P_{pv}	0	P_{batt}	Mode 2
0	0	1	0	P_{batt}	P_{pv}	Mode 3
0	1	0	P_{pv}	0	P_{pv}	Mode 4
0	1	0	P_{pv}	0	P_{pv}	Mode 5
0	1	1	P_{pv}	P_{batt}	$P_{pv} + P_{batt}$	Mode 6
1	0	0	P_{pv}	0	P_{pv}	Mode 7
1	1	0	P_{pv}	0	P_{pv}	Mode 8

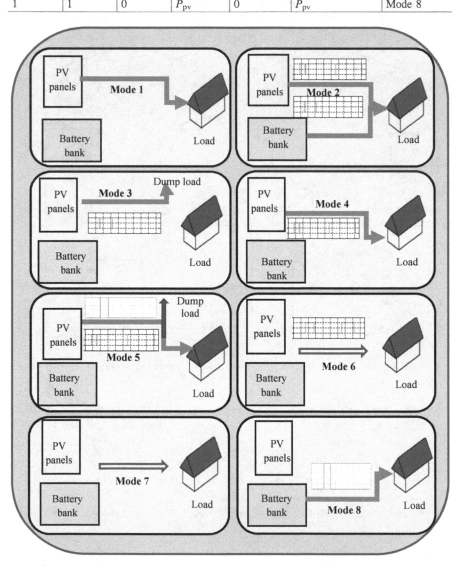

Fig. 6.13 Schematic power flow of PV/batteries system (second configuration)

6.4 Photovoltaic/Fuel Cell System

6.4.1 Power Management of Photovoltaic/Fuel Cell System

To supervise the PV/FC system (Fig. 6.14), the net power ΔP of the hybrid system is first calculated and then compared to the difference between the load power and the power produced by the PV source [10, 14, 33, 35].

$$P_{\text{Load}} = P_{\text{PV}} + P_{\text{FC}} \tag{6.1}$$

$$\Delta P = P_{\text{PV}} - P_{\text{Load}} \tag{6.2}$$

It is assumed that the photovoltaic generator is the main source and is sized to supply the load.

The different modes that control the operation of the proposed system are as follows (Fig. 6.15):

Mode 1: The load is supplied only by PV panels.
Mode 2: Photovoltaic power is insufficient to supply the load, so FC power is added to compensate.
Mode 3: Fuel cells are supplied the load.
Mode 4: If the net power is positive, the load is supplied by PV, and the electrolyzer is charging.

Table 6.3 represents different modes that can be distinguished during this operation.

6.4.2 Application Under MATLAB/Simulink

To make the supervision of this structure (PV/FC system), a precise sizing has been made (see Chap. 5) to obtain the PV panels number, fuel cell number based of course on a daily energy (D_{energy}).

Application: It has taken $N_{\text{pv}} = 20$ panels of 80 Wp, $D_{\text{energy}} = 2808$ Wh/day.

Using the same load profile given in Fig. 6.6, the obtained results are as follows (Figs. 6.16 and 6.17).

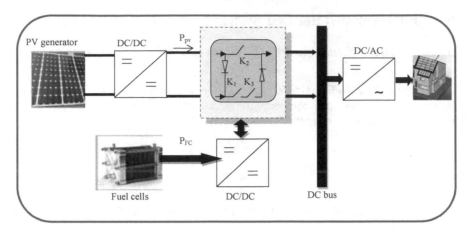

Fig. 6.14 Power management of PV/FC system

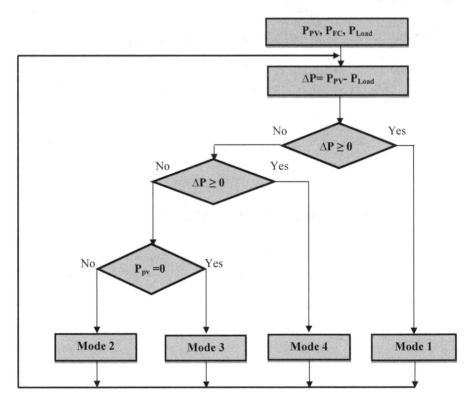

Fig. 6.15 Flowchart of the PV/FC system

Table 6.3 Switches state operating modes in the second structure

Modes	K_1	K_2	K_3	P_{load}
Mode 1	On	On	Off	P_{pv}
Mode 2	On	Off	On	$P_{pv} + P_{FC}$
Mode 3	Off	Off	On	P_{FC}
Mode 4	On	Off	Off	P_{pv}
Mode 5	Off	Off	Off	Disconnection

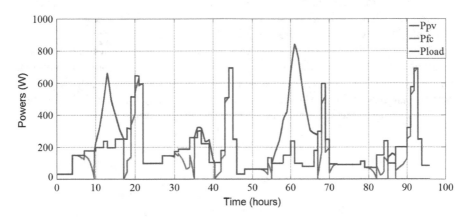

Fig. 6.16 Different power waveforms

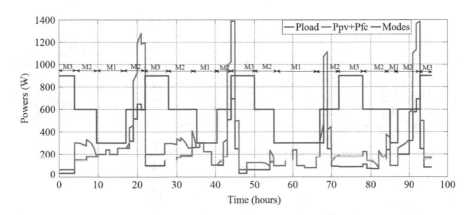

Fig. 6.17 Sum of the different source powers, the load power and the different modes

6.5   Photovoltaic/Battery/Fuel Cell System

6.5.1   Power Management of Photovoltaic/Battery/Fuel Cell System

In this system, PV panels, batteries and fuel cell energy systems integrated with an electrolyzer for hydrogen production have been used to supply a residential house (Fig. 6.18).

The power management control operates for charging batteries or for charging electrolyzer when there is PV power surplus and uses the FCs when there is no available PV power. In this structure, according to the available power value, there are three cases. If it is positive, PV panels will supply the load, while batteries are charging and if there is an excess of power it will be dissipated through a dump load. Hence, the needs of adding the fifth switch K_5. In the second case, the load is fed only by batteries, and in the last case, the load is supplied by PV panels. According to the different tests, the system operates in one of the following modes (Fig. 6.19).

The different modes depend on the five switches K_1, K_2, K_3, K_4 and K_5 (Table 6.4).

The PV energy is considered the main source. The batteries and the electrolyzer are recharged whenever there is enough energy to maintain their lifespan. They are only used if necessary, respecting the state of charge of the management control. Depending on variable metrological conditions and load profile variations, six modes are possible:

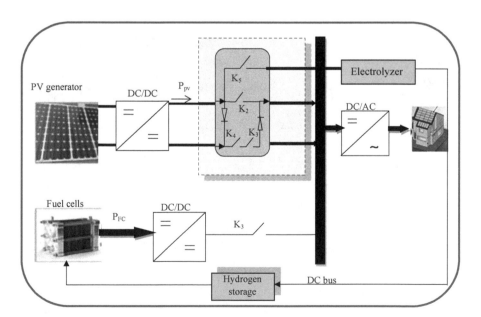

Fig. 6.18 Photovoltaic/battery/fuel cell system

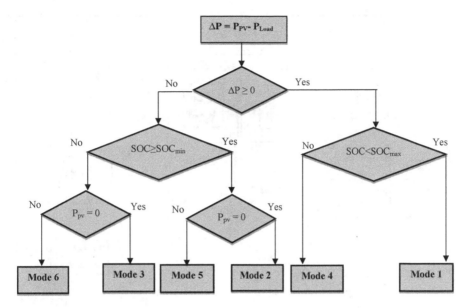

Fig. 6.19 Flowchart of the third structure

Table 6.4 Switches state operating modes in PV/batteries/FCs

K_1	K_2	K_3	K_4	K_5	P_{load}	Modes
1	0	0	0	0	P_{pv}	Mode 1
1	0	0	1	1	P_{pv}	Mode 2
1	0	0	0	1	P_{pv}	Mode 3
1	0	0	1	0	P_{pv}	Mode 4
1	1	1	0	0	$P_{\text{pv}} + P_{\text{batt}} + P_{\text{FC}}$	Mode 5
0	1	1	0	0	$P_{\text{batt}} + P_{\text{FC}}$	Mode 6

Mode 1: The load is supplied only by PV panels.

Mode 2: The load is supplied only by PV panels. The batteries will be charged, and the hydrogen tank will be also charged.

Mode 3: The load is supplied only by PV panels. The hydrogen tank is charged, so the batteries will be charged.

Mode 4: The load is supplied only by PV panels. The batteries are charged, so the hydrogen tank will be charged.

Mode 5: The photovoltaic generator cannot satisfy the load, and FC and battery powers are added to compensate PV panels.

Mode 6: The load is supplying only by FCs and batteries.

The different possible modes of operation are represented as in Fig. 6.20.

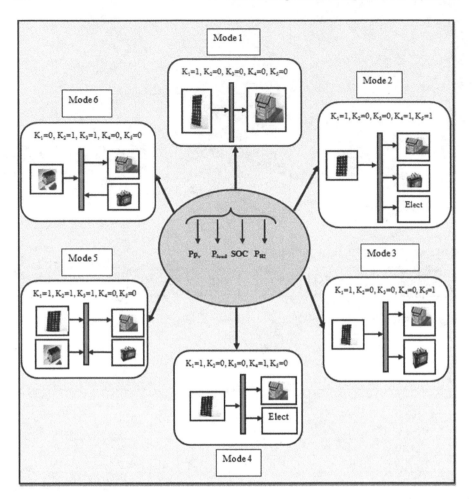

Fig. 6.20 Different modes of operation of the hybrid system studied

6.5.2 *Application Under MATLAB/Simulink*

To make the supervision of this structure (PV/battery/FC system), a precise sizing has been made (see Chap. 5) to obtain the PV panel number, battery number and fuel cell number which are based on a daily energy (D_{energy}).

Application: It has been taken $N_{pv} = 20$ panels of 80 Wp, $N_{batt} = 6$ batteries of 48 V, 200 Ah, $D_{energy} = 2808$ Wh/day.

Using the same load profile given in Fig. 6.6, the obtained results are as follows (Figs. 6.21 and 6.22). The sum of these powers is greater than the load power. From

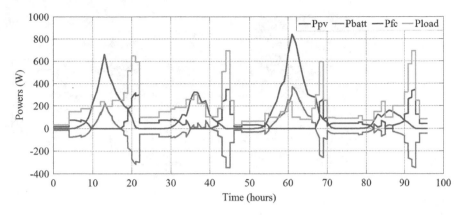

Fig. 6.21 Power variations of the different sources

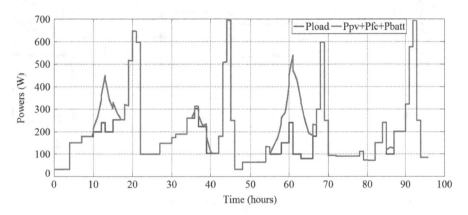

Fig. 6.22 Sum of the different source powers and the load power

these results, it can be concluded that the proposed management method work properly. In addition, there has been an excess that can be used to supply other additional loads.

Then, the different modes can be determinated as (Fig. 6.23).

6.6   Wind/Battery System

6.6.1   Power Management of Wind/Battery System

A power control management (PMC) of a hybrid wind/battery system can be proposed (Fig. 6.24). It determines the various operating processes of the hybrid system according to the weather conditions.

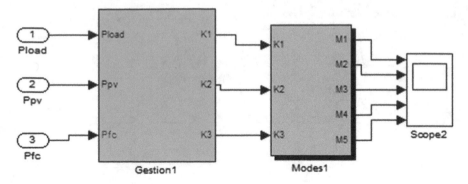

Fig. 6.23 Different modes under MATLAB/Simulink

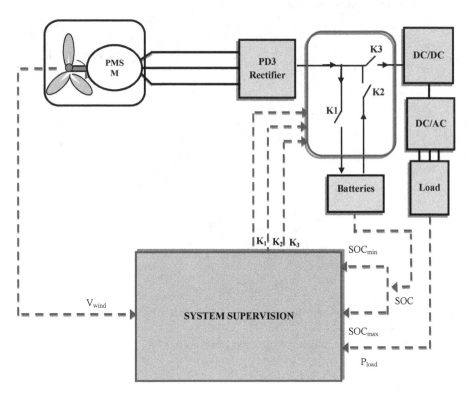

Fig. 6.24 Overall block diagram of the studied system

From the management algorithm, six different operating modes can be identified. The different powers involved are: the power supplied by the wind generator (P_{wind}), the power supplied or required by the battery for compensation or recharge, respectively (P_{batt}), the power dissipated (P_{dis}) and the power required by the load (P_{load}). The different modes are defined as follows (Table 6.5).

Table 6.5 Different modes in wind turbine/battery system

Modes	K_1	K_2	K_3	K_4	P_{load}
Mode 1	0	0	0	0	0
Mode 2	0	0	0	1	P_{wind}
Mode 3	0	0	1	1	$P_{wind} + P_{batt}$
Mode 4	0	1	0	1	$P_{wind} - \Delta P$
Mode 5	1	0	0	0	$P_{load} = 0\, P_{batt} = -P_{wind}$
Mode 6	1	0	0	1	$P_{wind} - P_{batt}$

Where

Mode 1: When the wind speed is below a minimum speed value, which represents the turbine start speed, the system stops and the load is disconnected to preserve the power stored by the storage.

Mode 2: When the power supplied by the generator is equal to the power required by the load, this one is supplied only by the wind generator power.

Mode 3: When the power supplied by the generator is insufficient to supply alone the load, energy compensation is essential. The lack is provided by the storage according to the state of charge.

Mode 4: When the power supplied by the generator is greater than the power required by the load, and the state of charge of the battery is greater than the maximum value, the battery is disconnected.

Mode 5: When the power supplied by the generator is less than the power required by the load, and the state of charge of the battery is less than the minimum value, the load is disconnected and the power supplied by the generator is directed send to the batteries for charging.

Mode 6: When the power supplied by the generator is greater than the power required by the load, the excess energy is directed send to the battery for charging as long as the state of charge is below the maximum value.

6.6.2 Application Under MATLAB/Simulink

To make the supervision of this structure (wind turbine/battery system), a precise sizing has been made (see Chap. 5) to obtain the wind turbine number and battery number based of course on a daily energy (D_{energy}).

Application: $N_{Wtb} = 1$ wind turbine of 1 kW (based on PMSM), $N_{batt} = 2$ batteries of 12 V, 110 Ah; it is considered that the batteries initially charged, $SOC_{min} = 30\%$, $SOC_{max} = 90\%$ and the initial SOC at 90%.

The studied system is developed using MATLAB®-SIMULINK® package. The Simulink model for the hybrid system is shown in Fig. 6.25.

The mathematical model of the supervision is represented in MATLAB/Simulink as in Fig. 6.26.

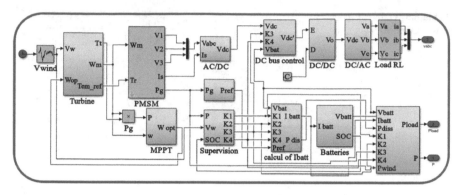

Fig. 6.25 Simulink model for PMC of wind turbine/battery configuration

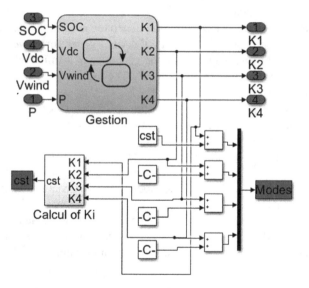

Fig. 6.26 Supervision model

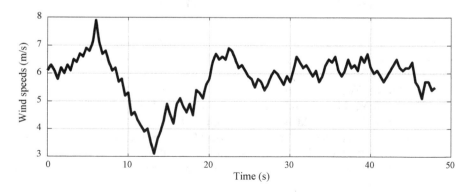

Fig. 6.27 Wind speed profile

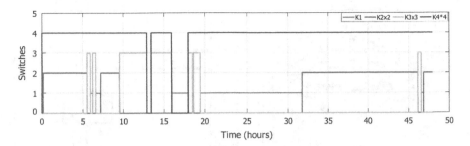

Fig. 6.28 Different switches K_1, K_2, K_3 and K_4

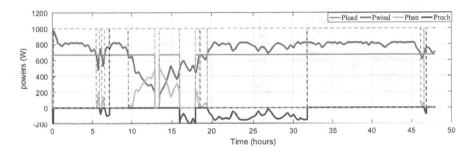

Fig. 6.29 Power waveforms during the two different days

Some simulation results show the performances of this hybrid structure (Figs. 6.27, 6.28 and 6.29).

6.7 Photovoltaic/Wind Turbine/Battery Structure

6.7.1 Power Management of Photovoltaic/Wind Turbine/ Batteries

The hybrid PV/wind turbine/battery system is represented as in Fig. 6.30. It consists of a photovoltaic system and wind power system associated with battery storage supplying centrifugal pump based on induction motor. To manage the different power flux, a supervisor control has been added. The use of the supervision allows producing maximum power from the PV and wind generators, protects the batteries against overcharge and depth discharge and satisfies the energy needs [19, 22].

The main objective of power management is in first part to satisfy the load power demand and in second part to maintain the state of charge of the battery bank to prevent blackout and to extend the battery life. The key decision factors for the

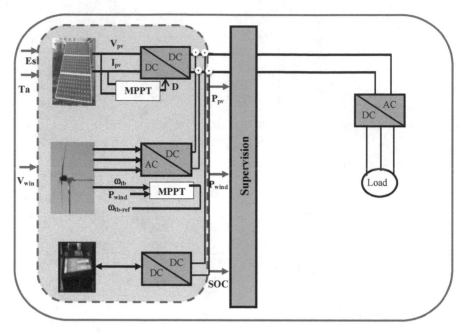

Fig. 6.30 Proposed PV/wind/batteries for standalone system

power management strategies are the power level provided by the photovoltaic generator, the wind generator and the state of charge (SOC) of the batteries. From this algorithm, we can establish the operation modes of the system management (Fig. 6.31):

$$P_{\text{hyb}} = P_{\text{pv}} + P_{\text{wind}} \tag{6.3}$$

- **Mode 1**: The power supplied by the PV arrays, and wind turbine (hybrid power) is equal to the load power.
- **Mode 2**: The hybrid power is less than the load power, so the lack of power will be compensated by the batteries, as the battery state of charge is superior to its minimum value.
- **Mode 3**: The hybrid power is insufficient to supply the load and the battery charge is less than its minimum value, so the battery is disconnected, and since the load demand is not satisfied, it will also be disconnected from this happens the battery connect again to store little energy available.
- **Mode 4**: When the available power exceeds the power required by the load, and that the battery charge is less than SOC_{max}, it supplies the load and charges the battery.

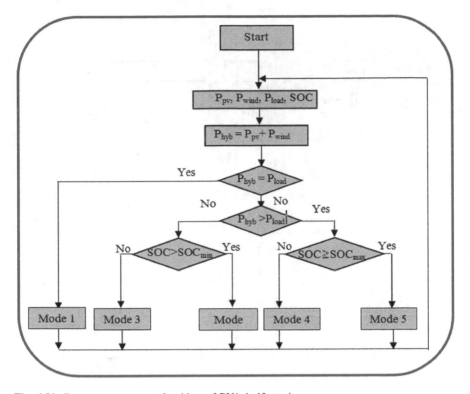

Fig. 6.31 Power management algorithm of PV/wind/batteries

- **Mode 5**: When the hybrid power exceeds the power required by the charge and that the battery charge is superior to SOC_{max}, disconnecting the battery and the load will be powered by the power supplied, and surplus energy will be directed to a dump load.

6.7.2   Power Management of Photovoltaic/Wind Turbine/ Batteries for Pumping Water

The system consists of a photovoltaic system and wind power system associated with battery storage supplying centrifugal pump based on induction motor (Fig. 6.32).

To control the DC bus voltage, field-oriented control (FOC) strategy can be applied to induction motor (IM) supplied by a hybrid system (Fig. 6.33). The reference speed is calculated from a reference power that varies according to the water flow rate (Fig. 6.34).

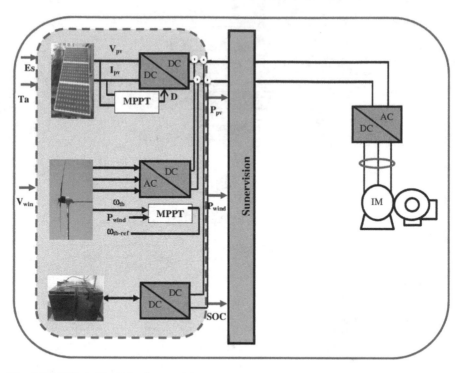

Fig. 6.32 PV/wind/batteries for standalone system

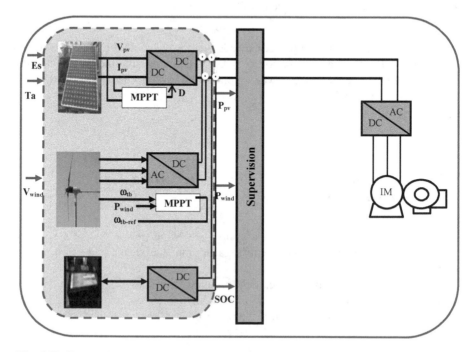

Fig. 6.33 Proposed PV/wind/batteries for pumping water system

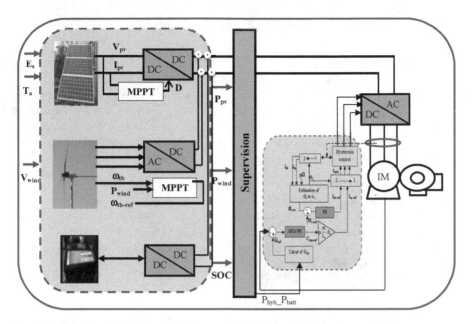

Fig. 6.34 Control of DC bus with FOC in hybrid pumping system

The output power P_{out} will be supplied induction moto-pump:

$$P_{out} = P_{hyb} + P_{batt}$$
$$P_{hyb} = P_{PV} + P_{wind}$$
(6.4)

$$T_{em} = K \cdot \omega^2$$
(6.5)

The mechanical output power of induction motor pump is:

$$P_{mec} = K \cdot \omega^3$$
(6.6)

where K is the pump constant

$$\omega = \sqrt[3]{P_{mec}/K}$$
(6.7)

The electromechanical torque of induction motor pump can be written as:

$$T_{em} = \sqrt[3]{K \cdot P_{mec}^2}$$
(6.8)

6.7.3 Application Under MATLAB/Simulink

The sizing of the various components of the PV/wind turbine/battery system has been made to satisfy the domestic needs of a family (see Chap. 5).

Application: N_{pv} = 12 panels of 175 Wc, N_{Wtb} = 1 kW, N_{batt} = 12 batteries of 100 Ah, N_{jaut}: number of days of autonomy N_{jaut} = 2 days, depth of discharge DOD = 0.8, battery efficiency η_{batt} = 0.9, a water tank of 80 m³, the dynamic level head is H = 10 m, the nominal flow rate is Q = 23 m³/h, an induction motor of 1.5 kW and the daily electrical energy required by the load is 5775 Wh/day.

Simulation Results
Simulation results are performed under varying environmental conditions to test the capacity of the global system to give the desired water flow according to user needs. The block diagram of studied system is illustrated in Fig. 6.35.

Some simulation results showing performances of this hybrid structure are given in Figs. 6.36, 6.37 and 6.38.

6.8 Photovoltaic/Wind Turbine/Battery/Diesel Generator System

6.8.1 Power Management of Photovoltaic/Wind Turbine/ Battery/Diesel Generator System

The management strategy for this hybrid energy system is to satisfy the load under variable weather conditions and manage the power flow while maintaining the

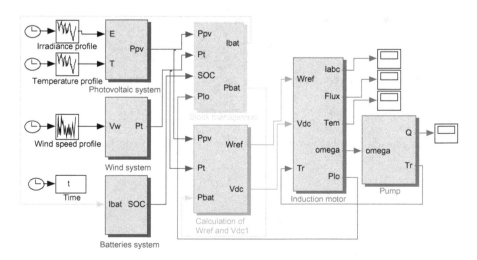

Fig. 6.35 Simulink block diagram of PV/wind/battery system supplying a water pump

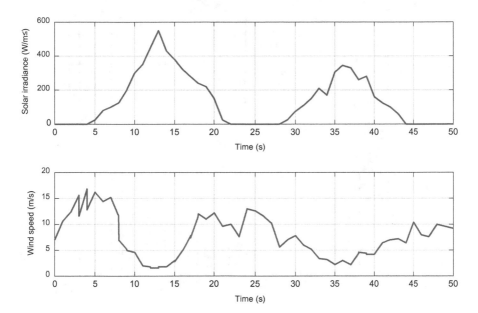

Fig. 6.36 Solar radiation and wind speeds

efficient operation of the various sources. The management strategy should mainly use the power generated by the PV and wind system to satisfy load. To supervise this system, fuzzy logic technique (FLC) is used (Fig. 6.39). The FLC principle consists of generating three control signals K_{pv}, K_{wind} and K_{DG}, from three inputs: solar irradiation G, wind speed V_{wind} and the battery state of charge (SOC). Fuzzy inference using the Mamdani method has been used, while defuzzification uses the center of gravity method [6] to calculate fuzzy logic outputs (k_{pv}, k_{wind} and k_{DG}). Where k_{pv} is the relay control signal of the photovoltaic generator, k_{wind} is the relay control signal of the wind generator and k_{DG} is the relay control signal of the diesel generator [22].

Table 6.6 shows the rule table of fuzzy controller.

Where the different values E_s, V_{wind} and SOC are as follows (Table 6.7).

The flowchart of the proposed PMC is as in Fig. 6.40 and Table 6.8.

$$P_{load} + P_{batt} = P_{pv} + P_{wind} + P_{DG} \tag{6.9}$$

Renewable energy sources (solar and wind) are the main used sources in the proposed hybrid system. The batteries are charged whenever there is enough energy, and in order to maintain their life longer, they will be used only if necessary while respecting their state of charge between SOC_{min} and a SOC_{max}. This is achieved through the proposed power management controller which works intelligently and precisely due to the FLC. According to random weather data and load profile variations, eight modes can be possible:

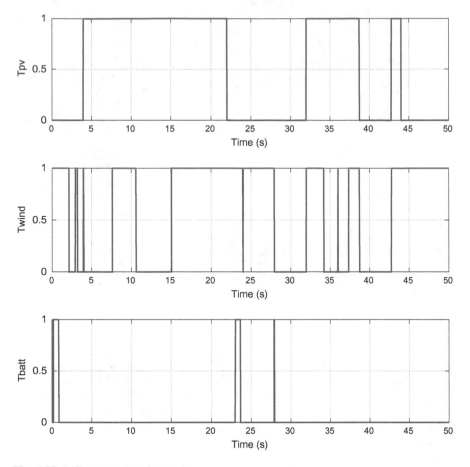

Fig. 6.37 Different modes of operation

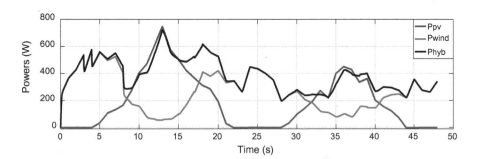

Fig. 6.38 Hybrid, battery and load power

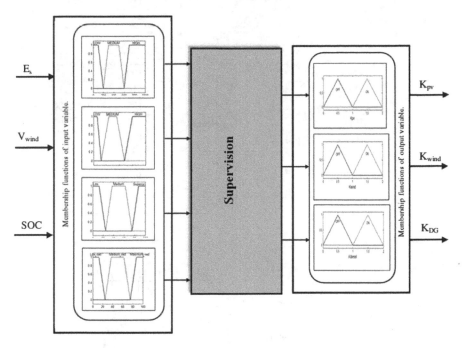

Fig. 6.39 Diagram of fuzzy controller for hybrid photovoltaic/wind/diesel system

Table 6.6 Fuzzy inference of fuzzy controller inputs/outputs

SOC	E_s	V_{wind}	P_{Load}		
			LOW	MED	MAX
LOW	LOW	LOW	101	101	101
	LOW	MED	110	110	110
	LOW	MAX	101	101	101
	MED	LOW	100	101	101
	MED	MED	110	110	110
	MED	MAX	100	101	101
	MAX	LOW	100	100	101
	MAX	MED	110	110	110
	MAX	MAX	100	100	101
MED	LOW	LOW	100	100	100
	LOW	MED	110	110	110
	LOW	MAX	100	100	100
	MED	LOW	100	100	100
	MED	MED	110	110	110
	MED	MAX	100	100	100
	MAX	LOW	100	100	100
	MAX	MED	110	110	110
	MAX	MAX	100	100	100

(continued)

Table 6.6 (continued)

SOC	E_s	V_{wind}	P_{Load}		
			LOW	MED	MAX
MAX	LOW	LOW	100	100	100
	LOW	MED	110	110	110
	LOW	MAX	100	100	100
	MED	LOW	100	100	100
	MED	MED	110	110	110
	MED	MAX	100	100	100
	MAX	LOW	100	100	100
	MAX	MED	110	110	110
	MAX	MAX	100	100	100

Table 6.7 Different values of E_s, V_{wind} and SOC

E_s (W/m^2)	0–200	200–600	600–100	
	LOW	MED	MAX	
V_{wind}(m/s)	0–3	3–12	12–20	
	LOW	MED	MAX	
SOC (%)	0–25	25–75	75–99	100
	LOW	MED	SUP	MAX

- **Mode 1**: In this case, the power of the three generators is sufficient to supply the load and the excess one is used to charge batteries.
- **Mode 2**: Wind turbine and PV panels are the main sources used. So both PV and wind contribute to supply the load. And of course, the excess if any is used to charge batteries.
- **Mode 3**: If the PV power is sufficient to supply the load, the charging is done by the PV panels. In general, this case occurs during a shiny day (summer).
- **Mode 4**: The PV source can be insufficient to supply load, so diesel generators are used.
- **Mode 5**: The wind turbine is the only source; it is the case of a winter day (no solar irradiation) or during night with high wind speeds.
- **Mode 6**: This case occurs when wind power is insufficient, and thus, the diesel generators start and compensate the power deficit.
- **Mode 7**: Only diesel generators supply the load.
- **Mode 8**: No sources are available to supply the load.

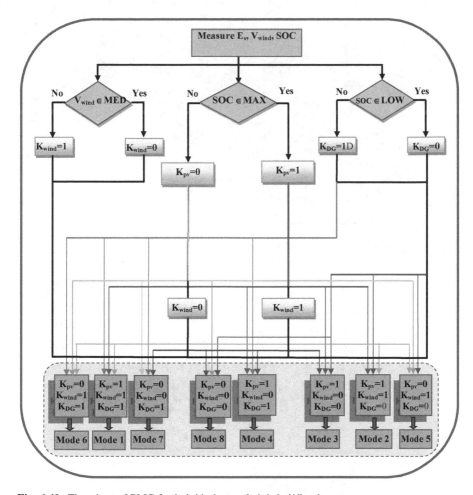

Fig. 6.40 Flowchart of PMC for hybrid photovoltaic/wind/diesel system

Table 6.8 Operation modes for hybrid photovoltaic/wind/diesel generator system

K_{pv}	K_{wind}	K_{DG}	P_{load}	Modes
1	1	1	$P_{load} = P_{pv} + P_{wind} + P_{DG} - P_{batt}$	Mode 1
1	1	0	$P_{Load} = P_{pv} + P_{wind} - P_{batt}$	Mode 2
1	0	0	$P_{Load} = P_{pv} - P_{batt}$	Mode 3
1	0	1	$P_{Load} = P_{pv} + P_{DG} - P_{batt}$	Mode 4
0	1	0	$P_{Load} = P_{Wind} - P_{batt}$	Mode 5
0	1	1	$P_{Load} = P_{wind} + P_{DG} - P_{batt}$	Mode 6
0	0	1	$P_{load} = P_{DG}$	Mode 7
0	0	0	$P_{Load} = -P_{batt}$ or $P_{Load} = 0$	Mode 8

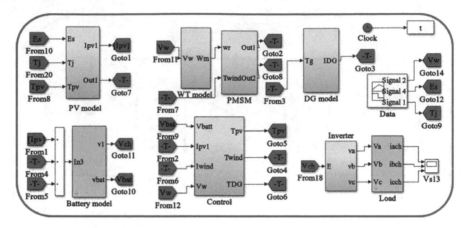

Fig. 6.41 Block system of the hybrid PV/wind/diesel generator system with battery storage

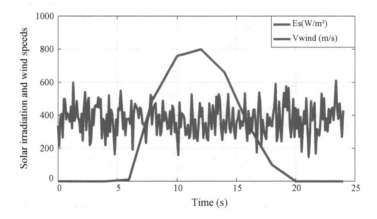

Fig. 6.42 Profile of the solar irradiation and wind speeds

6.8.2 Application Under MATLAB/Simulink

The block diagram under MATLAB/Simulink is represented in Fig. 6.41.

Application: The sizing of the various components of the PV/wind turbine/ batteries/diesel generator system has been made to satisfy a power load (see Chap. 5).

N_{pv} = 10 panels of 80 Wc, N_{Wtb} = 1 of 600 W, P_{DG} = 2 kVA, N_{batt} = 04 batteries of 192 Ah–12 V, N_{jaut}: number of days of autonomy = 2 days, depth of discharge DOD = 0.8, battery efficiency = 0.9.

Some simulation results show the performances of this hybrid structure (Figs. 6.42, 6.43 and 6.44).

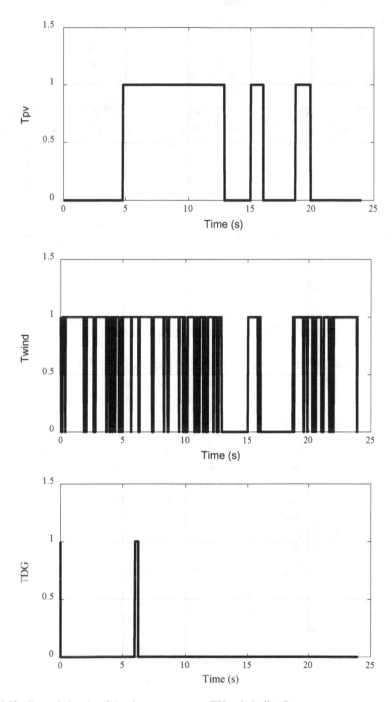

Fig. 6.43 Control signals of the three generators (PV, wind, diesel)

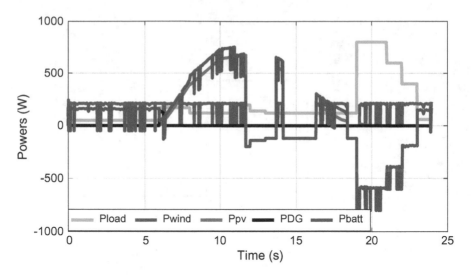

Fig. 6.44 Power waveforms

6.9   Photovoltaic/Wind Turbine/Batteries/Fuel Cells

6.9.1   Power Management of Photovoltaic/Wind Turbine/ Batteries/Fuel Cells

The management of the following structure is based on an algorithm that allows the supervision system of a hybrid system to decide how many and which generators to operate first, which loads are connected and how to use the storage (Fig. 6.45). This strategy makes it possible to improve the energy balance and reduce the use of FCs, which generate hydrogen consumption and also affect its lifespan. With this strategy, the FCs are switched off until the state of charge of the storage system reaches a minimum level.

This method must also take into account: prevent deep battery discharges, prevent battery overcharging and disconnection of the charge in the case of insufficient production and deep battery discharges.

Seven operating modes are then possible to determine the capacity of the hybrid system to meet the total power required (the loading power plus the power needed to charge the batteries), according to the weather conditions (solar irradiation, temperature and wind speed).

$$P_{\text{Hyb}} = P_{\text{PV}} + P_{\text{wind}} \tag{10}$$

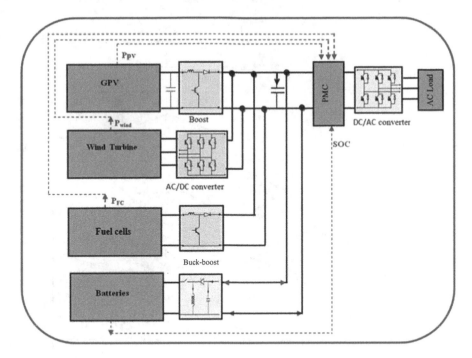

Fig. 6.45 PV/wind turbine/FC/battery hybrid structure with PMC

Mode 1: $P_{hyb} = P_{load}$, $P_{batt} = 0$

The power produced by the two generators (photovoltaic and wind) is sufficient to drive the motor pump.

Mode 2: $P_{hyb} > P_{load}$ and SOC $\geq$ SOC$_{max}$

The power produced by the two generators is sufficient to drive the load and the excess will be dissipated in a derivative load.

Mode 3: $P_{hyb} > P_{load}$ and SOC $<$ SOC$_{max}$

In this mode, the power produced by the two generators will supply the load, and the excess will be dissipated in a derivative load.

Mode 4: $P_{hyb} < P_{load}$ and SOC $>$ SOC$_{min}$

In this case, the power produced by the two generators is not sufficient to supply the charge, so batteries will be solicited to compensate the power generated by the photovoltaic and wind generator and at the same time satisfy the total load power.

Mode 5: SOC $\leq$ SOC$_{min}$ and $P_{hyb} + P_{FC} = P_{load}$

In this mode, since the batteries are discharged, the fuel cell is used to satisfy the load power required.

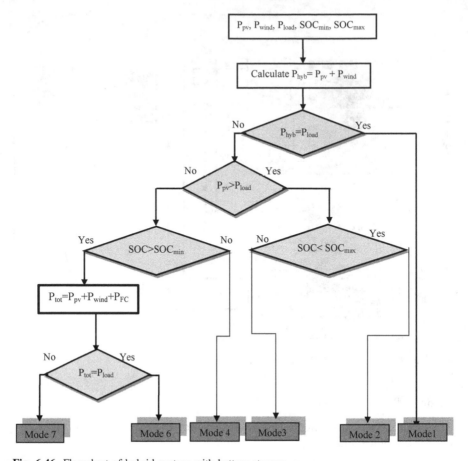

Fig. 6.46 Flowchart of hybrid system with battery storage

Mode 6: SOC $\leq$ SOC$_{min}$ and P_{hyb} + P_{FC} > P_{load},

In this mode, the load power is provided by the three generators (PV, wind turbine, FCs), and the surplus is used to charge batteries.

Mode 7: SOC $\leq$ SOC$_{min}$ and P_{hyb} + P_{FC} < P_{load},

In this mode, the charge is disconnected and the batteries are charged.
    The flowchart of this algorithm is as follows (Fig. 6.46).

6.9.2   Application Under MATLAB/Simulink

The sizing of the various components of the PV/wind turbine/FC/battery system has been made to satisfy the domestic needs of a family (see Chap. 5).

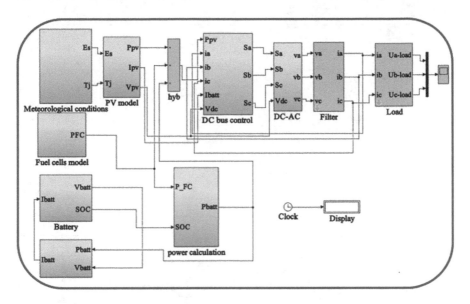

Fig. 6.47 Simulink block diagram of PV/wind/battery system supplying a water pump

Application: $N_{pv} = 12$ panels of 110 W_p, a wind turbine of 1 kW based on PMSG, $N_{batt} = 12$ batteries of 12 V, 100 Ah.

The block diagram under MATLAB/Simulink is represented in Fig. 6.47 with some simulation results showing performances of this hybrid structure (Figs. 6.48, 6.49 and 6.50).

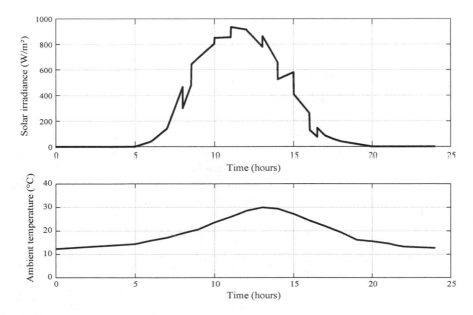

Fig. 6.48 Solar irradiance and temperature profile

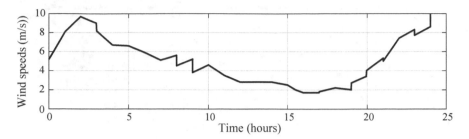

Fig. 6.49 Wind turbine profile

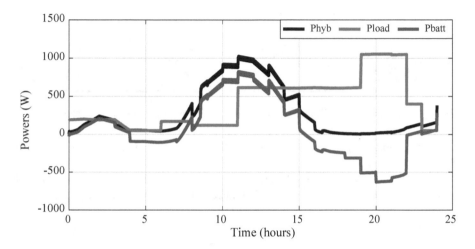

Fig. 6.50 Power variations of the different sources

6.10   PV System with Hybrid Storage (Batteries/Super-Capacitors)

This hybrid system is used to produce energy without interruption, and it consists of a photovoltaic generator (PV) and hybrid storage. In this case, it is considered batteries and super-capacitors (Fig. 6.51). As it has been explained in Chap. 4, this solution to combine batteries and super-capacitors is to cumulate their advantages (power and energy). This case is generally used for electric vehicles.

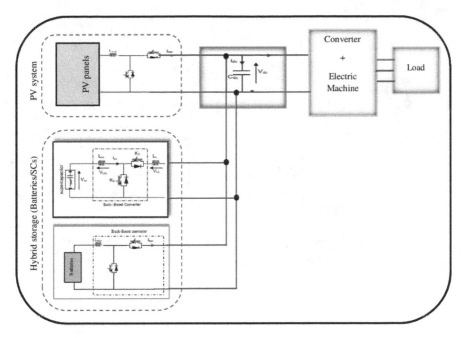

Fig. 6.51 PV system with hybrid storage (batteries/super-capacitors)

6.10.1 Control of DC Bus

To adapt the output voltage of the battery bank to the DC bus voltage, a DC/DC boost converter is used. The control is ensured with a proportional integral (PI) by regulation of the battery's current (Fig. 6.52). Two controllers proportional integral (PI) are also used to ensure the regulation of the SC's currents (Fig. 6.53).

The proposed system is represented as in Fig. 6.54.

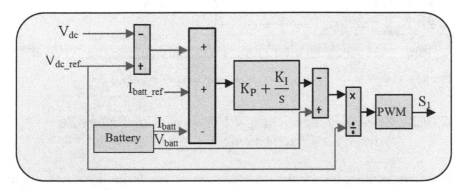

Fig. 6.52 Control of the DC/DC boost converter of battery

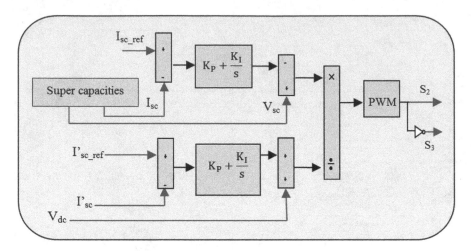

Fig. 6.53 Control of the DC/DC boost of super-capacitors

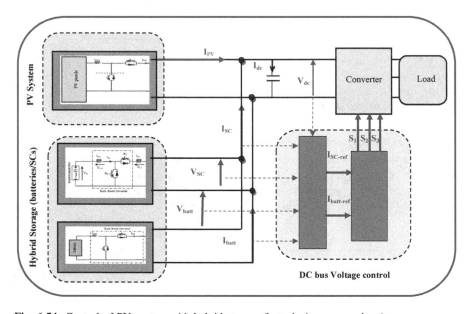

Fig. 6.54 Control of PV system with hybrid storage (batteries/super-capacitors)

6.10.2   Power Management of PV System with Battery/SC Storage for Electric Vehicle

In electric vehicles, using batteries as the main storage device is not sufficient to satisfy the power demand especially during acceleration and at sudden needs of power. Super-capacitors (SC) are generally used during rapid power changes and recover braking energy to improve the electric vehicle autonomy, while batteries are used to satisfy energy demand for a longer period of time [64]. The hybrid storage system is used to decrease battery stresses, their span life and of course which deals to improve the efficiency of the hybrid system. Power management control is necessary to ensure coordination between the different energy sources (Fig. 6.55).

Before sizing this hybrid system, it is needed to know the energy necessary for the vehicle operation. The different forces acting on a vehicle along a slope are (Fig. 6.56):

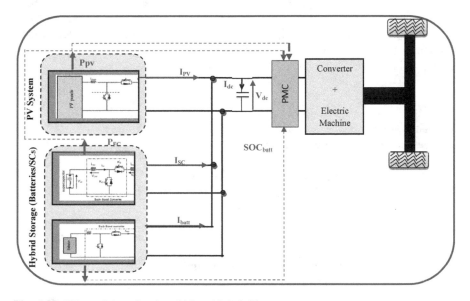

Fig. 6.55 PV supplying electric vehicle with hybrid storage

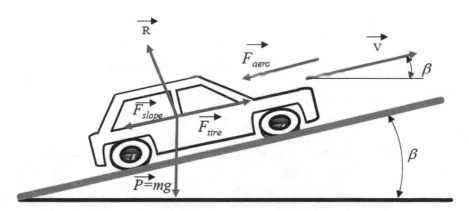

Fig. 6.56 Different forces acting on a vehicle moving along a slope

– Rolling resistance force F_{tire} due to the friction of the vehicle tires on the road. It is given as:

$$F_{tire} = m_{EV} \times g \times f_{ro} \tag{6.11}$$

With: m_{EV} is the vehicle total mass, g is the gravity acceleration, f_{ro} is the rolling resistance force constant.

– Aerodynamic drag force F_{aero} due to the friction on the body moving through the air. Its expression is:

$$F_{aero} = (1/2) \times \rho_{air} \times A_f \times C_d \times V^2 \tag{6.12}$$

With: ρ_{air} is the air density, A_f is the frontal surface area of the vehicle, C_d is the aerodynamic drag coefficient, V is the electric vehicle speed.

– Climbing force F_{slope} which depends on the road slope.

$$F_{slope} = m \times g \times \sin(\beta) \tag{6.13}$$

With: β is the road slope angle.
The total resistive force R is given as:

$$R = F_{tire} + F_{aero} + F_{slope} \tag{6.14}$$

The load torque T_r can be written as:

$$T_r = F_r \times r_{tire} \tag{6.15}$$

With: r_{tire} is the tire radius, F_r is the total force.
The total power is calculated as:

$$P_{load} = P_{pv} + P_{batt} + P_{SC} \tag{6.16}$$

6.11 PV/FCs with Hybrid Storage (Batteries/SCs)

6.11.1 Supervision of PV/FCs with Hybrid Storage (Batteries/SCs)

It consists of hybrid PV/FCs (Fig. 6.57). The principal source is PV system associated to hybrid storage (batteries/SCs).

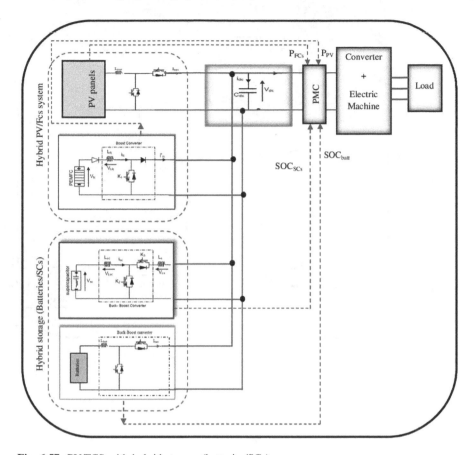

Fig. 6.57 PV/FCS with hybrid storage (batteries/SCs)

The flowchart of the supervision of hybrid storage associates to PV source is represented as in Fig. 6.58.

When the FCs are added, the flowchart will be Fig. 6.59.

6.12 Power Management of Fuel Cell/Battery System Supplying Electric Vehicle

Electric vehicle (EV) using battery storage must be recharged regularly. Those using fuel cells for supplying electrical energy, a supply for hydrogen is necessary. Generally, EV uses batteries for storage, but due to the less autonomy, hydrogen or fuel cell vehicle, solar vehicle or a combination of solar, FC and battery bank can be a competitive solution.

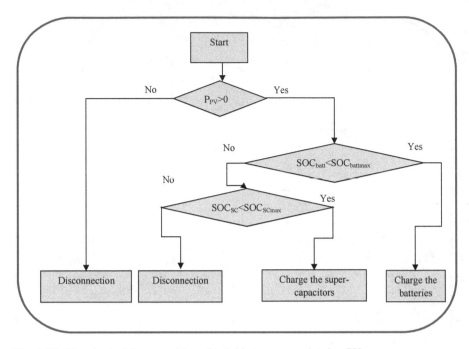

Fig. 6.58 Flowchart of the supervision of hybrid storage associated to PV source

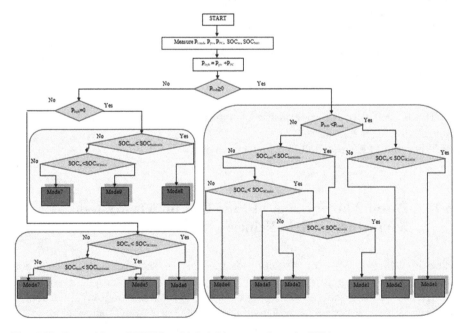

Fig. 6.59 Supervision of PV/FCs with hybrid storage (batteries/SCs)

6.12.1   First Structure

The system includes a fuel cell (PEMFC) as a main source and batteries (Fig. 6.60). The DC bus is connected to a voltage inverter to supply the motor via the PMC system to make coordination between the different energy sources.

The energy management system is made in accordance with vehicle dynamics. It uses a simple algorithm to choose at any time the best power flow between the different energy sources in order to minimize fuel consumption and satisfy the load needs. Three modes of vehicle operation are possible: stop mode, traction mode and braking mode (Fig. 6.61; Table 6.9).

$$P_{\text{load}} = P_{\text{batt}} + P_{\text{FC}} \tag{6.17}$$

$$\Delta P_{\text{load}} = P_{\text{FC}} - P_{\text{Load}} \tag{6.18}$$

where ΔP_{load} is the variation of the power demand required by the electric vehicle.

The flowchart of this algorithm is shown in Fig. 6.61.

The flowchart of power management of fuel cell/battery system supplying electric vehicle is given as (Fig. 6.62).

The different modes depend on the three switches K_1, K_2 and K_3 (Table 6.10).

6.12.2   Second Structure

This strategy allows us to recover the energy supplied by fuel cells in case of surplus energy when batteries are fully charged (Fig. 6.63). This energy can be stored in super-capacitor or used to generate hydrogen in the electrolyzer.

The different modes are as follows (Fig. 6.64):

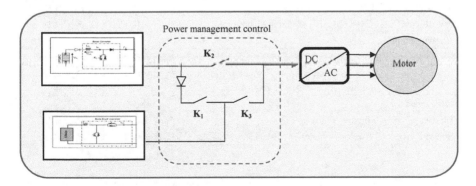

Fig. 6.60 Supervision of FC/battery system supplying electric vehicle: first structure

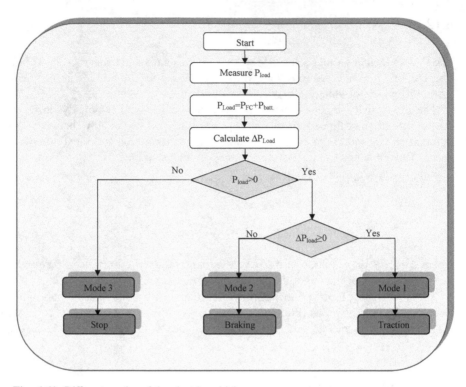

Fig. 6.61 Different modes of the electric vehicle

Table 6.9 Modes of vehicle operation

Cases		Powers
First case	FCs supply only EV	$P_{load} > 0$, $P_{FC} > 0$ and $P_{batt} = 0$
Second case	FCs and batteries supply together EV	$P_{load} > 0$, $P_{FC} > 0$, $P_{batt} > 0$
Third case	Batteries supply only EV	$P_{load} > 0$, $P_{FC} = 0$, $P_{batt} > 0$
Fourth case	FCs supply EV and reload batteries	$P_{load} > 0$, $P_{FC} > 0$, $P_{batt} < 0$
Braking mode		
First case	No energy flow	$P_{load} = 0$, $P_{FC} = 0$, $P_{batt} = 0$
Second case	Batteries recover energy braking kinetic and also receive power from FCs.	$P_{load} < 0$, $P_{FC} > 0$, $P_{batt} < 0$
Stop mode		
First case	No energy flow	$P_{load} = 0$, $P_{FC} = 0$, $P_{batt} = 0$
Second case	FC reloads batteries	$P_{load} = 0$, $P_{FC} > 0$, $P_{batt} < 0$

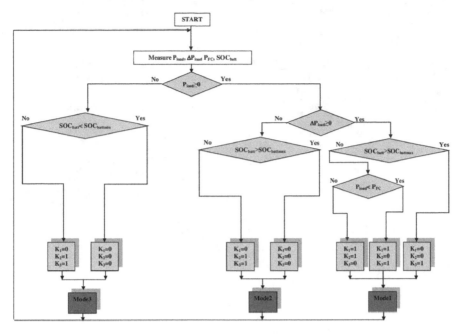

Fig. 6.62 Flowchart of power management of fuel cells/batteries system supplying electric vehicle (first structure)

- **Mode 1**: The FC power produced is upper than the power of load ($P_{FC} > P_{Load}$) and the SOC of the batteries is less than the SOC_{max}, in this case, the excess energy is stored in batteries.
- **Mode 2**: The power supplied by FC is insufficient ($0 < P_{FC} < P_{load}$); in this case, the power of batteries is added to satisfy the power demand. It is the compensation mode.
- **Mode 3**: The FC power is sufficient as $P_{FC} = P_{Load}$.
- **Mode 4**: This mode is operating when no energy provides from the FC, so batteries supply alone the load.
- **Mode 5**: The produced FC power is insufficient to supply the load, and the battery is discharged ($SOC < SOC_{min}$); in this case, the batteries charge.
- **Mode 6**: In this case, there is no FC energy production and batteries are discharged, the load is disconnected.
- **Mode 7**: In this mode, the FC power ($P_{FC} > P_{Load}$) is quite sufficient to supply the load, and the excess energy will be stored in the auxiliary source.
- **Mode 8**: The load is disconnected ($P_{Load} = 0$) and the $SOC = SOC_{max}$ (battery charged), so the fuel cell energy is stored in the auxiliary source.

The different modes depend on the four switches K_1, K_2, K_3 and K_4. It is shown in Table 6.11.

Table 6.10 Simplified table in the case of three switches

Switches			Powers			SOC	Vehicle's state	Modes
K_1	K_2	K_3	P_{FC}	P_{batt}	P_{Load}			
0	0	0	0	0	0	$=SOC_{min}$	Stop	3
0	0	0	0	0	$\Delta P_{Load} < 0$	$=SOC_{max}$	Braking	2
1	0	0	$0 < P_{FC} < P_{Load}$	P_{FC}	0	$<SOC_{max}$	Stop	3
1	0	0			$\Delta P_{Load} < 0$		Braking	2
0	1	0	$P_{FC} \geq P_{Load}$	0	$P_{Load} = P_{FC}$	$>SOC_{min}$	Traction	1
1	1	0	$P_{FC} > P_{Load}$	$P_{FC} - P_{Load}$	$P_{Load} = P_{FC} - P_{Load} > 0$	$<SOC_{max}$	Traction	1
0	0	1	0	P_{batt}	$P_{Load} = P_{batt} > 0$	$>SOC_{min}$	Traction	1
0	1	1	$0 < P_{FC} < P_{Load}$	P_{batt}	$P_{Load} = P_{batt} + P_{FC} > 0$	$>SOC_{min}$	Traction	1

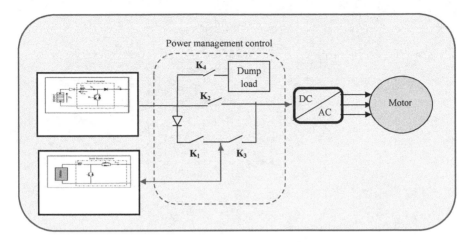

Fig. 6.63 Supervision of FC/battery system supplying electric vehicle: second structure

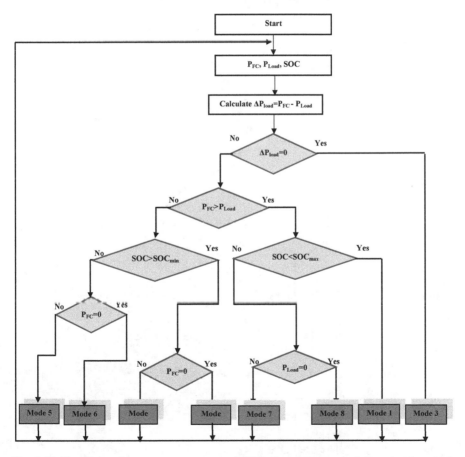

Fig. 6.64 Flowchart of power management of FC/battery system supplying electric vehicle (second structure)

Table 6.11 Table in the second structure

Switches				Powers				SOC	Vehicle's state	Modes
K_1	K_2	K_3	K_4	P_{FC}	P_{batt}	P_{dump}	P_{Load}			
0	0	0	0	0	0	0	0	$=SOC_{min}$	Stop	Mode 6
0	0	0	1	P_{FC}	0	P_{FC}	0	$=SOC_{max}$	Stop	Mode 8
							$\Delta P_{Load} < 0$		Braking	
0	0	1	0	0	$P_{batt} - P_{Load}$	0	P_{batt}	$>SOC_{min}$	Traction	Mode 4
0	1	0	0	$P_{FC} = P_{Load}$	0	0	P_{FC}	$=SOC_{max}$	Traction	Mode 3
0	1	0	1	$P_{FC} > P_{Load}$	0	$P_{FC} - P_{Load}$	P_{FC}	$=SOC_{max}$	Traction	Mode 7
0	1	1	0	$0 < P_{FC} < P_{Load}$	P_{batt}	0	$P_{FC} + P_{batt}$	$>SOC_{min}$	Traction	Mode 2
1	0	0	0	P_{FC}	P_{FC}	0	0	$<SOC_{min}$	Stop	Mode 5
							$\Delta P_{Load} < 0$		Braking	
1	1	0	0	P_{FC}	$P_{FC} - P_{Load}$	0	P_{FC}	$<SOC_{max}$	Traction	Mode 1

6.13 Conclusions

Through this chapter, we have tried to show that the power management strategy depends on the type of energy system and its components. Different structure has been presented, and some examples under MATLAB/Simulink have been given in order to better understand these algorithms.

References

1. Ally CZ, Sun Y, De Jong ECW (2018) Impact of virtual inertia on increasing the hosting capacity of island diesel-PV ac-grid. In: 2018 53rd international universities power engineering conference (UPEC 2018), Art. no. 8541859
2. Rezkallah M, Singh S, Chandra A (2017) Real-time hardware testing, control and performance analysis of hybrid cost-effective wind-PV-diesel standalone power generation system. In: 2017 IEEE Industry Applications Society Annual Meeting (IAS 2017)
3. Rezkallah M, Chandra A, Saad M, Tremblay M, Singh B, Singh S, Ibrahim H (2018) Composite control strategy for a PV-wind-diesel based off-grid power generation system supplying unbalanced non-linear loads. In: 2018 IEEE Industry Applications Society Annual Meeting (IAS 2018), 8544618
4. Jamalaiah A, Raju CP, Srinivasarao R (2017) Optimization and operation of a renewable energy based pv-fc-micro grid using homer. In: Proceedings of the international conference on inventive communication and computational technologies (ICICCT 2017), Art. no. 7975238, pp 450–455. https://doi.org/10.1109/icicct.2017.7975238
5. Ntziachristos L, Kouridis C, Samaras Z, Pattas K (2005) A wind-power fuel-cell hybrid system study on the non-interconnected Aegean islands grid. Renew Energy 30(10):1471–1487. https://doi.org/10.1016/j.renene.2004.11.007
6. Gharibi M, Askarzadeh A (2019) Size optimization of an off-grid hybrid system composed of photovoltaic and diesel generator subject to load variation factor. J Energy Storage 25:100814
7. Hatti M, Meharrar A, Tioursi M (2011) Power management strategy in the alternative energy photovoltaic/PEM fuel cell hybrid system. Renew Sustain Energy Rev 15(9):5104–5110. https://doi.org/10.1016/j.rser.2011.07.046
8. Marzband M, Azarinejadian F, Savaghebi M, Guerrero JM (2017) An optimal energy management system for islanded microgrids based on multiperiod artificial bee colony combined with Markov chain. IEEE Syst J 11(3):1712–1722, Art. no. 7101215. https://doi.org/10.1109/jsyst.2015.2422253. http://www.ieee.org/products/onlinepubs/news/0806_01.html
9. Rekioua D (2018) Energy management for PV installations. Adv Renew Energies Power Technol 1:349–369
10. Mokrani Z, Rekioua D, Mebarki N, Rekioua T, Bacha S (2017) Energy management of battery-PEM Fuel cells Hybrid energy storage system for electric vehicle. In: Proceedings of 2016 international renewable and sustainable energy conference (IRSEC 2016), Art. no. 7984073, pp 985–990
11. Miron C, Christov N, Olteanu SC (2016) Energy management of photovoltaic systems using fuel cells. In: 2016 20th international conference on system theory, control and computing (ICSTCC 2016)—joint conference of SINTES 20, SACCS 16, SIMSIS 20—proceedings 7790757, pp 749–754
12. Sami BS, Sihem N, Bassam Z (2018) Design and implementation of an intelligent home energy management system: a realistic autonomous hybrid system using energy storage. Int J Hydrogen Energy 43(42):19352–19365

13. Zaouche F, Rekioua D, Gaubert J-P, Mokrani Z (2017) Supervision and control strategy for photovoltaic generators with battery storage. Int J Hydrogen Energy 42(30):19536–19555
14. Mokrani Z, Rekioua D, Mebarki N, Rekioua T, Bacha S (2017) Proposed energy management strategy in electric vehicle for recovering power excess produced by fuel cells. Int J Hydrogen Energy 42(30):19556–19575
15. Lazizi A, Kesraoui M, Achour D, Chaib A (2017) Power control and management of PV/battery/diesel/water pumping system. In: 2017 8th International Renewable Energy Congress, (IREC 2017), Art. no. 7925999
16. Lopez-Flores DR, Duran-Gomez JL (2018) Control and energy management system techniques in renewable sources: a brief review. In: International Power Electronics Congress (CIEP), Art. no. 8573320, pp 139–145, Oct 2018
17. Naik BB, Rambabu M (2019) Energy management by using renewable energy sources. Int J Innov Technol Exploring Eng 8(8):315–322
18. Poompavai T, Kowsalya M (2019) Control and energy management strategies applied for solar photovoltaic and wind energy fed water pumping system: a review. Renew Sustain Energy Rev 108–122
19. Serir C, Rekioua D, Mezzai N, Bacha S (2016) Supervisor control and optimization of multi-sources pumping system with battery storage. Int J Hydrogen Energy 41(45):20974–20986
20. Dahbi S, Aziz A, Messaoudi A, Mazozi I, Kassmi K, Benazzi N (2018) Management of excess energy in a photovoltaic/grid system by production of clean hydrogen. Int J Hydrogen Energy 43(10):5283–5299
21. Lingamuthu R, Mariappan R (2019) Power flow control of grid connected hybrid renewable energy system using hybrid controller with pumped storage. Int J Hydrogen Energy 44(7):3790–3802
22. Roumila Z, Rekioua D, Rekioua T (2017) Energy management based fuzzy logic controller of hybrid system wind/photovoltaic/diesel with storage battery. Int J Hydrogen Energy 42(30):19525–19535
23. Riveron I, Gomez JF, Gonzalez B, Mendez JA (2019) An intelligent strategy for hybrid energy system management. Renew Energy Power Quality J 17:550–554
24. Chaouali H, Salem WB, Mezghani D, Mami A (2018) Fuzzy logic optimization of a centralized energy management strategy for a hybrid PV/PEMFC system feeding a water pumping station. Int J Renew Energy Res 8(4):2190–2198
25. Bhuyan SK, Hota PK, Panda B (2018) Modeling, control and power management strategy of a grid connected hybrid energy system. Int J Electr Comput Eng 8(3):1345–1356
26. Bendary AF, Ismail MM (2019) Battery charge management for hybrid PV/wind/fuel cell with storage battery. Energy Procedia 162:107–116
27. Zakzouk NE, Dyasty AE, Ahmed A, El Safty SM (2018) Power flow control of a standalone photovoltaic-fuel cell-battery hybrid system. In: 7th International IEEE conference on renewable energy research and applications (ICRERA 2018), pp. 431–436, 8566869
28. Abdelkafi A, Masmoudi A, Krichen L (2018) Assisted power management of a stand-alone renewable multi-source system. Energy 145:195–205
29. Halima NB, Ammous S, Oualha A (2018) Constant-power management algorithms of a hybrid wind energy system. J Renew Sustain Energy 10(5):055102
30. Gam O, Abdelati R, Abdou Tankari M. Mimouni MF (2019) An improved energy management and control strategy for wind water pumpingsystem. Trans Inst Measure Control (in press)
31. Zaouche F, Mokrani Z, Rekioua D (2017) Control and energy management of photovoltaic pumping system with battery storage. In: Proceedings of 2016 international renewable and sustainable energy conference (IRSEC 2016), pp. 917–922, Art. no. 7983890
32. Zaouche F, Rekioua D, Mokrani Z (2018) Power flow management for stand alone PV system with batteries under two scenarios. In: Proceedings of 2017 international renewable and sustainable energy conference (IRSEC 2017), 8477328

33. Mokrani Z, Rekioua D, Rekioua T (2014) Modeling, control and power management of hybrid photovoltaic fuel cells with battery bank supplying electric vehicle. Int J Hydrogen Energy 39(27):15178–15187

34. Jiang H, Xu L, Li J, Hu Z, Ouyang M (2019) Energy management and component sizing for a fuel cell/battery/supercapacitor hybrid powertrain based on two-dimensional optimization algorithms. Energy 386–396

35. Mebarki N, Rekioua T, Mokrani Z, Rekioua D (2015) Supervisor control for stand-alone photovoltaic/hydrogen/ battery bank system to supply energy to an electric vehicle. Int J Hydrogen Energy 40(39):13777–13788

36. Hajizadeh A, Golkar MA (2007) Intelligent power management strategy of hybrid distributed generation system. Electr Power Energy Syst 29:783–795

37. Kamal T, Hassan SZ (2016) Energy management and simulation of photovoltaic/hydrogen/ battery hybrid power system. ASTES J 1(2):11–18

38. Singh SN, Snehlata (2017) Intelligent home energy management by fuzzy adaptive control model for solar (pv)-grid/dg power system in India. Int J Power Control Signal Comput (IJPCSC) 2(2):61–66

39. Singh SN, Singh AK (2012) Rural home energy management by fuzzy control model for solar (PV)-grid/ DG power system in India. J Electr Control Eng (JECE) 2(1):29–33

40. Rabhi A, Bosch J, Elhajjaji A (2015) Energy management for an autonomous renewable energy system. Energy Procedia 83:299–309

41. Koohi-Kamali S, Rahim NA, Mokhlis H (2014) Smart power management algorithm in microgrid consisting of photovoltaic, diesel, and battery storage plants considering variations in sunlight, temperature, and load. Energy Convers Manage 84:562–582

42. Al-Norya M, El-Beltagya M (2014) An energy management approach for renewable energy integration with power generation and water desalination. Renew Energy 72:377–385

43. Miceli R (2013) Energy management and smart grids. Energies 6:2262–2290

44. Sudipta C, Manoja DW, Simoes MG (2007) Distributed intelligent energy management system for a single-phase high-frequency AC micro grid. IEEE Trans Ind Electron 54(1):97–109

45. Courtecuisse V, Sprooten J, Robyns B, Petit M, Francois B, Deuse J (2010) A methodology to design a fuzzy logic based supervision of hybrid renewable energy systems. Math Comput Simul 81(2):208–224

46. Jeong KS, Lee WY, Kim CS (2005) Energy management strategies of a fuelcell/battery hybrid system using fuzzy logic. J Power Sources 145(2):319–326

47. Karami N, Moubayed N, Outbib R (2014) Energy management for a PEMFC–PV hybrid system. Energy Convers Manage 82:154–168

48. Ipsakis D, Voutetakis S, Seferlis P, Stergiopoulos F, Elmasides C (2009) Power management strategies for a stand-alone power system using renewable energy sources and hydrogen storage. Int J Hydrogen Energy 34(16):7081–7095

49. García P, Torreglosa JP, Fernández LM, Jurado F (2013) Optimal energy management system for stand-alone wind turbine/photovoltaic/hydrogen/battery hybrid system with supervisory control based on fuzzy logic. Int J Hydrogen Energy 38(33):14146–14158

50. Andari W, Ghozzi S, Allagui H, Mami A (2017) Design, modeling and energy management of a PEM fuel cell/supercapacitor hybrid vehicle

51. Pinto PJR, Rangel CM (2010) A power management strategy for a stand-alone photovoltaic/ fuel cell energy system for a 1 kW application. In: Hydrogen energy and sustainability—advances in fuel cells and hydrogen workshop 3rd seminar international Torres Vedras (Portugal), 29–30 April 2010

52. Jaganmohan Reddy Y, Pavan Kumar YV, Padma Raju K, Ramsesh A (2013) PLC based energy management and control design for an alternative energy power system with improved power quality. Int J Eng Res Appl (IJERA) 3(3):186–193

53. Belvedere B, Bianchi M, Borghetti A, Nucci CA, Paolone M, Peretto A (2012) A microcontroller-based power management system for standalone micro grids with hybrid power supply. IEEE Trans Sustain Energy 3(3):422–431

54. Sekar K, Duraisamy V (2015) Efficient energy management system for integrated renewable power generation systems. J Sci Ind Res 74:325–329
55. Santos GV, de Oliveira FH, Cupertino AF, Pizziollo TA, Pereira HA (2013) Power flow management in hybrid power system using flatness based control. In: IEEE PES conference on innovative smart grid technologies (ISGT Latin America), pp 1–6, 15–17 April 2013
56. Wang C, Nehrir MH (2008) Power management of a stand-alone wind/photovoltaic/fuel cell energy system. IEEE Trans Energy Convers 23:957–967
57. Zhang J, Huang L, Shu J, Wang H, Ding J (2017) Energy management of PV-diesel-battery hybrid power system for island stand-alone micro-grid. Energy Procedia 105:2201–2206
58. Shaahid SM, El-Amin I (2009) Techno-economic evaluation of off-grid hybrid photovoltaic–diesel–battery power systems for rural electrification in Saudi Arabia—a way forward for sustainable development. Renew Sustain Energy Rev 13:625–633
59. Kaldellis J, Zafirakis D, Kavadias K, Kondili E (2012) Optimum PV-diesel hybrid systems for remote consumers of the Greek territory. Appl Energy 97:61–67
60. Yilmaz S, Dincer F (2017) Optimal design of hybrid PV-diesel-battery systems for isolated lands: a case study for Kilis, Turkey. Renew Sustain Energy Rev 77:344–352
61. Marzband M, Ghadimi M, Sumper A, Domínguez-García JL (2014) Experimental validation of a real-time energy management system using multi-period gravitational search algorithm for microgrids in islanded mode. Appl Energy 128:164–174
62. Trovão JPF, Santos VDN, Antunes CH, Pereirinha PG, Jorge HM (2015) A real-time energy management architecture for multisource electric vehicles. IEEE Trans. Ind. Electron. 62 (5):3223–3233, Art. no. 6975185. http://ieeexplore.ieee.org/xpl/tocresult.jsp?isnumber= 5410131. https://doi.org/10.1109/tie.2014.2376883
63. Rekioua D, Zaouche F, Hassani H, Rekioua T, Bacha S (2018) Modeling and fuzzy logic control of a stand-alone photovoltaic system with battery storage. Turk J Electromechanics Energy 4(1)
64. Rekioua D, Mokrani Z, Rekioua T (2018) Control of fuel cells-electric vehicle based on direct torque control. Turk J Electromechanics Energy 3(2)
65. Ye Y, Sharma R, Garg P (2014) An integrated power management strategy of hybrid energy storage for renewable application. In: IECON Proceedings (Industrial Electronics Conference), Art. no. 7048951, pp 3088–3093
66. Serpi A, Porru M, Damiano A (2017) An optimal power and energy management by hybrid energystorage systems in microgrids. Energies 10(11), Art. no. 1909
67. Olatomiwa L, Mekhilef S, Ismail MS, Moghavvemi M (2016) Energy management strategies in hybrid renewable energy systems: a review. Renew Sustain Energy Rev 62:821–835. https://doi.org/10.1016/j.rser2.2016.05.040

CPSIA information can be obtained
at www.ICGtesting.com
Printed in the USA
LVHW082105141220
674148LV00001B/26

9 783030 340230